Green Separation and Purification Technology

Green Separation and Purification Technology

Guest Editor

Shun Yao

MDPI

Basel • Beijing • Wuhan • Barcelona • Belgrade • Novi Sad • Cluj • Manchester

Guest Editor
Shun Yao
School of Chemical
Engineering
Sichuan University
Chengdu
China

Editorial Office
MDPI AG
Grosspeteranlage 5
4052 Basel, Switzerland

This is a reprint of the Special Issue, published open access by the journal *Separations* (ISSN 2297-8739), freely accessible at: https://www.mdpi.com/journal/separations/special_issues/M6952F43Q1.

For citation purposes, cite each article independently as indicated on the article page online and as indicated below:

Lastname, A.A.; Lastname, B.B. Article Title. *Journal Name* **Year**, *Volume Number*, Page Range.

ISBN 978-3-7258-7282-4 (Hbk)
ISBN 978-3-7258-7283-1 (PDF)
https://doi.org/10.3390/books978-3-7258-7283-1

Contents

About the Editor

Shun Yao

Shun Yao obtained his PhD in Pharmacy from China Pharmaceutical University in 2008. Since September 2008, he has worked at the School of Chemical Engineering at Sichuan University, serving as the Deputy Director of the Department of Pharmaceuticals and Bioengineering. Currently, he is the leader of the Green Pharmaceutical Engineering research team and a doctoral supervisor. He has long been engaged in research on the development of green pharmaceutical engineering, as well as related functional materials, their active ingredients, and technologies for their separation/analysis in the context of traditional Chinese medicine and natural medicines. In addition, he has conducted research on the comprehensive development and utilization of natural biomass resources, and on the application of computational chemistry and chiral technology in the field of biomedicine. He has led more than ten basic research projects, including projects supported by the National Natural Science Foundation and the Provincial Science and Technology Support Program, as well as Municipal Key R&D Projects. He has published over 300 academic papers in both Chinese and English, and has participated in the publication of more than 15 textbooks and monographs; most recently, he edited a monograph titled Natural Products and New Green Solvents, which was published by Chemical Industry Press (Beijing, China) in 2025 (ISBN: 9787122465214); it was supported by the National Natural Science Foundation (82273814), as is the case with this book. In addition, he has applied for and authorized over 20 invention patents, and his related achievements have been reported by the international Sci & Tech media platform Vertical News. Finally, he was awarded second prize for the National Science and Technology Progress Award and second prize for the Ministry of Education's Science and Technology Progress Award.

separations

Article

Strengthened Effect of Surface-Active Ionic Liquids on Curcumin Solubility and Extraction Performance of Curcuminoids

Dan Li [1], Yuxin Qin [1], Jingxing Li [1], Subhan Mahmood [1], Jianqin Shi [1], Yu Cao [2,*] and Shun Yao [1,*]

1 School of Chemical Engineering, Sichuan University, Chengdu 610065, China; 2023223070096@stu.scu.edu.cn (D.L.); 2023223075215@stu.scu.edu.cn (Y.Q.); linzi20022003@126.com (J.L.); engnr.chem.subhan@gmail.com (S.M.); sjq1232024@163.com (J.S.)

2 College of Life Science & Biotechnology, Mianyang Normal University, Mianyang 621000, China

* Correspondence: agnes_cy2004@163.com (Y.C.); cusack@scu.edu.cn (S.Y.); Tel.: +86-028-85405221 (S.Y.)

Abstract: As a kind of bioactive component in the rhizome of natural plant *Curcuma longa* L. (turmeric), curcumin is almost insoluble in water at neutral and acidic pH, which limits its further utilization and development. At the same time, traditional extraction and separation processes typically require the use of a large number of organic solvents. Ionic liquids (ILs) are organic molten salts with melting points below 100 °C. When an ionic liquid exists in a liquid state at or near room temperature, it is referred to as a room-temperature ionic liquid (RTIL). They have a temperature range, good physical and chemical stability, and good structural designability. They have a strong solubilization enhancement effect for many organic compounds. This study first explored the molecular forms of curcumin in ionic liquid aqueous solutions and the intermolecular interactions between curcumin and ionic liquids using spectral analysis and computational chemistry methods; furthermore, using an ionic liquid aqueous solution as an extraction agent, curcumin-like substances (curcuminoids) were extracted from turmeric powders under ultrasound assisted conditions, revealing the relationship between the structure of the ionic liquid and the extraction efficiency. After that, a kinetic study was conducted for the extraction of curcuminoids from turmeric powders, using second-order kinetics fitting to obtain the rate constant and initial extraction rate during the extraction process. Finally, the comparison with a ComplexGAPI tool and antioxidant experiment was performed on the extraction by using ionic liquids and traditional solvent. The full results can provide reference for the design of IL extractants and their application for natural products.

Keywords: surface-active ionic liquids; turmeric; critical micelle concentration; solubilization enhancement; extraction

1. Introduction

From the current perspective, the micelle aggregation behavior of surfactants is of great significance for insoluble substances, including naturally active products [1]. In aqueous solution, when the concentration of surfactants reaches a certain value, they will aggregate to form micelles. The concentration at which surfactants begin to aggregate is called the critical micelle concentration (CMC) [2]. As the concentration of surfactants increases, the morphology of micelle aggregates gradually presents as follows: (1) relatively small spherical micelles; (2) slender cylindrical rod-shaped micelles with hemispherical ends; (3) larger flat-layered micelles; (4) a vesicle containing multiple spherical structures, composed of bilayer lamellar micelles arranged in one or more concentric spheres [3]. The solubilization effect of micelles is an important property of surfactants. Although both soluble and insoluble substances in solvents can be dissolved through the mechanism of micelle solubilization, from a practical application perspective, it is more meaningful that micelles can dissolve substances in typically insoluble solvents [4]. The solubilization ability of micelles towards target molecules is also influenced by intermolecular interactions. In the study of solubilization of ibuprofen and gefitolozil, the anionic surfactant sodium dodecyl sulfate (SDS) has the lowest solubilization ability, while cationic and non-ionic surfactants have stronger solubilization ability. This is due to the dissociation of ibuprofen and gemfibrozil molecules into anions, which exhibit charge repulsion with the anionic head group of SDS [5]. In the solubilization of fenofibrate, danazol, and androstane, the solubility of danazol in ionic surfactants is much higher than that of fenofibrate and androstane. This is due to the ion–dipole interaction between polar danazol molecules and the charged head groups of ionic surfactants [6].

There are many factors that affect the solubilization of micelles, among which the most important are the structure of surfactants and solubilizers. As a kind of green solvent with excellent solubility, ionic liquids (ILs) exhibit good solubility for many organic and inorganic substances [7]. Those ILs with long carbon chains in cations often have properties similar to traditional surfactants, and those that can aggregate in water to form micelles can be called surface-active ionic liquids (SAILs) [8]. SAIL not only has good surface activity but also has a wide liquid temperature range and good "structural designability". It has a wide range of application prospects in the fields of compound extraction and separation, drug delivery carrier construction, and micelle catalysis [9,10]. Generally speaking, as the carbon chain length of ionic liquids increases, CMC decreases, while the aggregation number increases. Table 1 summarizes the reported CMC and N_{agg} number of imidazolium bromide ILs in previous studies [11–18]. Although there is a certain research foundation on the micelle aggregation behavior of ionic liquids, there are still few reports on the micelle solubilization of ionic liquids. Previous studies have suggested that ionic liquids can enhance the water solubility of some insoluble organic compounds and biomolecules. For instance, a non-linear correlation of the concentration of ionic liquid with [Cnmim]+ cation and casein solubility was observed with a significant maximum at a concentration of approximately 40% [19]. In another case, the obviously increased solubility of glibenclamide (Glib) resulted from the formation of complex hydrogen bonds and the pi–pi interaction of the aromatic rings with the IL [20]. As a hydrophobic natural product, its solubility can be improved by non-ionic surfactant micelles induced by ionic liquids, which resulted from the formation of small sized mixed micelles (hydrodynamic diameter $\approx$ 10 nm) at a high number density [21]. When ionic liquid surfactants are used for extraction, the target substance micelle distribution coefficient is the driving force for successfully extracting it into ionic liquid micelles. That is to say, the greater the partition coefficient of a component from the aqueous phase to the ionic liquid micelles, the higher its extractability. Generally speaking, hydrophobic substances (octanol–water partition coefficient greater than 300) are more suitable for extraction with ionic liquid surfactants than polar substances, which provides the possibility for selective extraction [22].

Curcumin (Cur) and curcuminoids (e.g., demethoxycurcumin, DMC; bisdemethoxy-curcumin, BDMC) are a kind of yellow pigments extracted from the rhizomes of ginger plants such as turmeric in the ginger family. Their structures belong to acidic polyphenols with an unsaturated aliphatic and aromatic main chain [23]. Curcumin is commonly used as a coloring agent and acid–base indicator for meat products and also has pharmacological effects such as anti-inflammatory and antioxidant properties [24,25]. It is an orange yellow crystalline powder with a slightly bitter taste, which is insoluble in water at neutral and acidic pH. This property limits its further utilization and development. In this study, curcumin and curcuminoids were chosen as the target molecules, and the solubilization enhancement effects of alkyl imidazolium bromine ionic liquids ([$C_n$mim]Br) were firstly explored through a series of experiments and computation for them. In addition, key conditions were investigated for the extraction of curcuminoids from turmeric powders, and classical kinetic models were applied to fit the whole process. Finally, a Complex-GAPI tool and an antioxidant experiment were performed for the comparison on the extraction by using ionic liquids and traditional solvent. The study was aimed to provide meaningful reference for the application of green extractants and assessment for natural products extraction.

Table 1. Currently reported CMC and N_{agg} number of imidazolium bromide ILs.

ILs	CMC (mM)	Aggregation Number
[C16mim]Br	0.55–0.84 [11], 0.8 [12], 0.61 [13], 0.653 [14]	64 [11], 66 [13], 76 [14]
[C14mim]Br	2.8–2.9 [11], 2.5 [13], 2.61 [14]	48 [11], 59 [13], 61 [14]
[C12mim]Br	10.9 [11,15], 4.3 [12], 11.2 [11], 9.8 [13], 9.53 [14], 11 [16], 8.5 [15]	37 [11], 44 [13], 43 [14]
[C10mim]Br	40 [17], 41 [13], 29.3–32.9 [15], 30–40 [18], 31 [16]	42 [13], 37–47 [18], 27 [16]
[C9mim]Br	74 [13]	45 [13]
[C8mim]Br	121 [12], 150 [18], 141 [16]	21–27 [18], 25 [16]
[C6mim]Br	470 [12], 400–600 [18], 797 [16]	-

2. Materials and Methods

2.1. Reagents and Materials

1-methylimidazole, 1-bromobutane, 1-bromohexane, 1-bromooctane, 1-bromodecane, 1-bromododecane, potassium hexafluorophosphate, sodium tetrafluoroborate, potassium hydrogen sulfate, anhydrous sodium acetate, dodecyltrimethylammonium bromide, methylene chloride, 1,1-diphenyl-2-picrylhydrazyl free radical (DPPH-D273092, 97%), and sodium lauryl sulfate were all purchased from Aladdin Chemical Reagent Company (Shanghai, China). Ultrapure (UP) water was obtained from a Milli-Q water purification system (Millipore, Bedford, MA, USA). The main ILs used in this study included 1-octyl-3-methylimidazolium bromine ([C_8mim]Br), 1-decyl-3-methylimidazolium bromine ([C_{10}mim]Br), 1-dodecyl-3-methylimidazolium bromine ([C_{12}mim]Br), N-octyl tropine bromine ([C_8Tr]Br), N-decyl tropine bromine ([C_{10}Tr]Br), and N-dodecyl tropine bromine ([C_{12}Tr]Br), which were synthesized according to the methods in the literature [26–28] (see Figure 1a,b), and their spectral purities of nuclear magnetic resonance (NMR) were above 96.5% (see Supplementary Materials). Curcumin and curcuminoids (99% purity) were purchased from MeilunBio (Dalian, China, see Figure 1c). Acetonitrile, ethyl acetate, dichloromethane, and anhydrous ethanol were purchased from Kelong chemical factory (Chengdu, China).

2.2. Instruments

A DF-101S oil bath pot and DZF-6050 vacuum-drying oven were procured from Yinyu-yuhua Instrument Co., Ltd. (Gongyi, China). The FA2004B electronic balance was sourced from Tianmei Balance Instrument Co., Ltd. (Shanghai, China). The SHZ-D (III) circulation vacuum pump was from Dufu Instrument Factory (Zhengzhou, China). The RE-2000 rotary evaporator was from Yarong Biochemical Instrument Factory (Shanghai, China). The DC-2006 low-temperature constant temperature bath was from Hengping

Instrument Factory (Shanghai, China). The 85-2 magnetic stirrer was from Sile Instrument Factory (Shanghai, China). The DDS-12A digital conductivity meter was from Hongyi Instrument Co., Ltd. (Shanghai, China). The AOE 380 UV–Vis Spectrometer was from Aoyi Instrument Co., Ltd. (Shanghai, China). The YM-031S ultrasonic generator was from Fangao Microelectronics Co., Ltd. (Shenzhen, China).

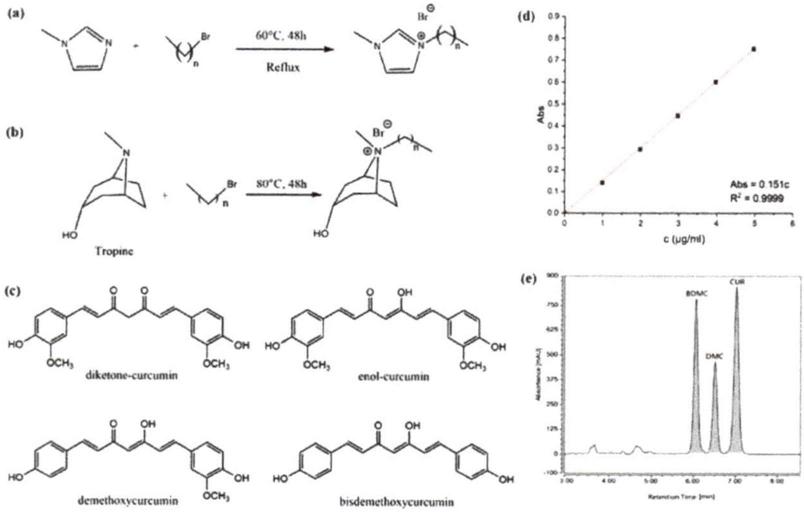

Figure 1. Synthesis routes of (**a**) imidazolium and (**b**) tropine-based ILs; (**c**) different structures of curcumin and related curcuminoids; (**d**) UV−Vis calibration curve of curcumin; and (**e**) HPLC (High Performance Liquid Chromatography) chromatogram of curcuminoids (CUR: curcumin, BDMC: bisdemethoxycurcumin, and DMC: demethoxycurcumin).

2.3. Determination of Critical Micelle Concentration of Imidazolium Ionic Liquids by Conductivity

According to the reported study [29], a certain amount of ionic liquid was accurately weighed to prepare a stock solution, which was then diluted to different concentrations of IL aqueous solutions. During measurement, 5 mL of the IL aqueous solution was taken and placed in a test tube. The probe of the conductivity meter was inserted into the test tube, ensuring that the electrode was fully submerged in the solution. After that, the test tube containing the probe was placed in a glass jacket, and the temperature was remained within the jacket at 25 °C using a constant temperature water bath circulator. Simultaneously, a magnetic stirrer was employed for agitation (300 rpm). Finally, the conductivity data were recorded once the values stabilized (see the details in Section S1 of Supplementary Materials).

2.4. UV–Vis Detection Method for Curcumin and Establishment of Its Standard Curve

In the study of enhanced solubility, UV–Vis was used for the quantitation of curcumin. The absorbance of IL exists in the far UV region, and the absorbance of curcumin was measured at visible region, which ensured the accurate measurement. On the basis of current method [30], 25.0 mg of curcumin was precisely weighed and dissolved in ethanol to prepare solutions with concentrations of solution with only ethanol added, 1.0 $\mu g \cdot mL^{-1}$, 2.0 $\mu g \cdot mL^{-1}$, 3.0 $\mu g \cdot mL^{-1}$, 4.0 $\mu g \cdot mL^{-1}$, and 5.0 $\mu g \cdot mL^{-1}$. Ethanol was treated as the reference solution, and the absorbance was measured at 425 nm. With the concentration of curcumin as the x-axis and the absorbance as the y-axis, linear fitting was performed on the UV–Vis absorbance data to obtain the quantitative standard curve for curcumin. As illustrated in Figure 1, within the concentration range of 0.0 μg·mL^{-1} to 5.0 μg·mL^{-1}, the absorbance of curcumin demonstrates a strong linear relationship with its concentration, with the linear equation being Abs = 0.151c and a correlation coefficient (R^2) of 0.9999 (see Figure 1d).

2.5. Establishing the Method for Determining Curcumin Solubility

Aqueous solutions of ionic liquid were prepared with concentrations of 10 mM, 50 mM, 100 mM, 300 mM, and 1000 mM; and 5 mL of each solution was placed in the separate test tubes. An excess amount of curcumin was added into each solution to ensure supersaturation. The system temperature was maintained at 25 °C using a thermostatic circulating water bath and the mixture was stirred with a magnetic stirrer. Stirred for 12 h to achieve dissolution equilibrium, the system was promptly sampled, and the solutions were filtered. The filtrate was then diluted with ethanol in appropriate ratios, followed by UV–Vis spectrophotometric analysis at a detection wavelength of 425 nm to determine the concentration of curcumin, thereby ascertaining its solubility in the various concentrations of ionic liquid aqueous solutions.

2.6. Extraction Conditions and Calculation of Extraction Yield

The details of the extraction experiment were introduced as in Section S2 of Supplementary Materials. Because of the complex composition of the extract, HPLC was used in the study of curcuminoid extraction from turmeric with ionic liquid. In the reverse-phase system under current HPLC conditions, the ionic liquid will not be effectively retained on the chromatographic column, which can be rapidly eluted as polar component and separated from the curcuminoids. Following the methods and conditions described in references [31,32], curcuminoids were analyzed using the following HPLC conditions: C_{18} reverse-phase chromatographic column (250 × 4.6 mm, 5 μm; Welch materials, Shanghai, China) as stationary phase; acetonitrile—0.5% (v/v) aqueous acetic acid solution (60:40) as mobile phase; 430 nm of detection wavelength; 0.9 mL·min^{-1} as flow rate; 20 μL of injection volume; 30 °C as column temperature. Figure 1e illustrates the HPLC chromatogram of the extracted curcuminoids, and it can be found that three compounds were well separated.

According to literature reports under identical HPLC conditions, the absorbance peaks at related retention time of 7.0 min, 6.5 min, and 6.1 min correspond to curcumin, demethoxycurcumin, and bisdemethoxycurcumin, respectively. The calculation of the extraction yield of curcuminoids is shown in Equations (1)–(3).

$$E_{CUR} = \frac{C_{CUR} \times V}{m} \times 100 \tag{1}$$

$$E_{DMC} = \frac{C_{DMC} \times V}{m} \times 100 \tag{2}$$

$$E_{BDMC} = \frac{C_{BDMC} \times V}{m} \times 100 \tag{3}$$

where E is the extraction yield of curcuminoids (g·100g^{-1}), C is the concentration of curcuminoids in the extract determined by HPLC (mg·mL^{-1}), V is the total volume of the extract (mL), and m is the mass of the turmeric powders (mg).

2.7. Extraction Kinetics Fitting

To elucidate the behavior of the mass transfer of curcuminoids from turmeric powders to the extract, the following simplifications and assumptions were made as follows:

(1) although the extraction process may involve various substances, only the extraction of curcuminoids is considered for simplicity; (2) the extraction process is assumed to commence with the diffusion of curcuminoids from the cell walls into the solution; (3) under identical extraction conditions, the concentration of curcuminoids at saturation is considered constant. The extraction process of curcuminoids can be modeled by a second-order kinetic equation (Equation (4)) [33,34].

$$\frac{dc_t}{dt} = k(C_s - C_t)^2 \tag{4}$$

where k is the extraction rate constant ($mL{\cdot}mg^{-1}min^{-1}$), C_S ($mg{\cdot}mL^{-1}$) represents the saturation concentration of curcuminoids at extraction equilibrium, signifying the maximum extraction capacity; and C_t ($mg{\cdot}mL^{-1}$) is the concentration of curcuminoids at any given time t (min). By separating variables in Equation (4), Equation (5) can be obtained. Integrating Equation (5) and applying the initial and boundary conditions: at $t = 0$, $C_t = 0$, and at $t = t$, $C_t = C_t$, allows Equation (6) to be obtained. Ultimately, the functional relationship between C_t and t is established as Equation (8).

$$\frac{dC_t}{(C_s - C_t)^2} = kdt \tag{5}$$

$$\frac{1}{C_s - C_t} - \frac{1}{c_s} = kt \tag{6}$$

$$C_t = C_s - \frac{C_s}{1 + C_s kt} \tag{7}$$

$$C_t = \frac{{C_s}^2 kt}{1 + C_s kt} \tag{8}$$

Transforming Equation (8) to Equation (9) yields a linear relationship between t/C_t and t. At $t = 0$, with $C_t = 0$, the initial extraction rate can be defined as Equation (11). In fitting the extraction kinetic data, t/C_t is used as the ordinate and t as the abscissa. Through linear fitting, C_S, h, and k can be calculated.

$$C_t = \frac{{C_s}^2 kt}{1 + C_s kt} \tag{9}$$

$$\frac{C_t}{t} = \frac{1}{\left(1/k{C_s}^2\right) + (t/C_s)} \tag{10}$$

$$h = k{C_s}^2 \tag{11}$$

$$C_t = \frac{t}{1/h + (t\,/\,C_s)} \tag{12}$$

2.8. Computational Chemistry Study

The structure optimization of ionic liquids and curcumin molecules was first carried out using Gaussian 09 W, and the selected optimization method was DFT B3LYP. Molecular electrostatic potential (ESP) is of great significance in investigating intermolecular electrostatic interactions, predicting reaction sites, and predicting molecular properties. It is widely used because molecules always approach each other in a complementary manner of electrostatic charges. ESP analysis was carried out on the van der Waals surface. A series of quantitative analyses of molecular van der Waals surface ESP was carried out with Multiwfn 3.6, using the 0.001 a.u. isoelectronic density surface as the van der Waals surface.

3. Results and Discussion

3.1. Critical Micelle Concentration of Imidazolium ILs in Aqueous Solutions

The methods for determining the critical micelle concentration (CMC) of surfactants include conductivity method, surface tension method, UV absorption method, fluorescence probe method, dynamic light scattering method, viscosity method, calorimetry method, and NMR method. The conductivity method for determining the CMC of ionic liquids is simple and accurate. When the concentration of ionic liquids with surface activity is lower than CMC, they are dispersed in water as monomers, and the conductivity of the solution shows a linear increase with the increase in ionic liquid concentration. Once the concentration reaches CMC, the added ionic liquid begins to associate to form micelles, and the degree of ionization of the micelle complex is usually lower than its monomer. At this point, the rate of increase in solution conductivity slows down, and the conductivity exhibits a "turning point" with the change of concentration. After performing segmented linear fitting on conductivity and concentration, the "inflection point" of the fitting corresponds to the CMC of the ionic liquid. The ratio of the slope after the "inflection point" to the slope before the "inflection point" is the ionization degree of the aggregate [35]. Here, Figure 2 shows the relationship between the conductivity and concentration of different types of imidazole ionic liquids in aqueous solutions. The longer the carbon chain of the ionic liquid, the smaller its CMC.

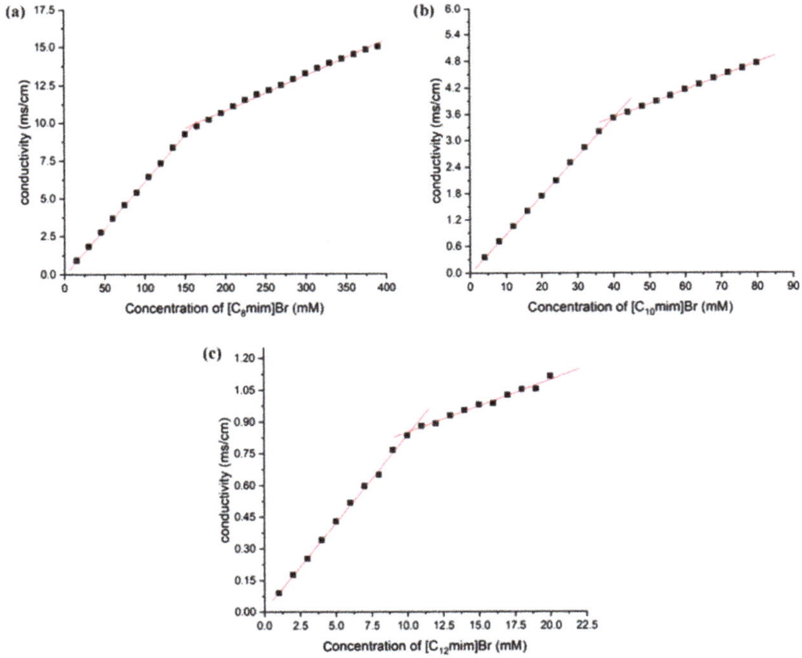

Figure 2. The relationship between conductivity and concentration of imidazolium ILs in aqueous solution (25 °C, (**a**) [C_8mim]Br, (**b**) [C_{10}mim]Br, and (**c**) [C_{12}mim]Br).

When the concentration is not too high, the aggregates of long-chain imidazolium ILs in water are spherical or nearly spherical, which are assumed to contain a hydrocarbon core consisting entirely of the hydrocarbon chain. The core of the micelle is composed of carbon chains on imidazole cations. According to the spherical volume formula, the volume of the micelle core ($N \bullet V_{core}$) is $4\pi L^3/3$. The micelle aggregation number of ionic liquids can be calculated by the following formula [36]:

$$N = \frac{4\pi L^3}{3V_{core}} \tag{13}$$

$$L = (1.5 + 1.265 n_{core}) \times 10^{-8} \tag{14}$$

$$V_{core} = (27.4 + 26.9 n_{core}) \times 10^{-24} \tag{15}$$

where L is the length of the alkyl chain embedded in the core of the micelle; n_{core} is the carbon number in the core. For ionic liquids [$C_n$mim] Br (n = 8, 10, and 12), $n_{core} \approx n_c - 1$, where n_c is the number of carbon atoms in the carbon chain [37]. The calculated aggregation number (N) and volume of single hydrocarbon core (V_{core}) and the length of the carbon chain in the micelle core (L) of imidazole ionic liquids [$C_n$mim]Br (n = 12, 10, 8) are shown in Figure 3. The longer the carbon chain of ionic liquids, the greater the aggregation number of micelles, and the larger the micelle size. Here, the solvent is water as the polar component, and long alkyl-chain-substituted ILs represent less-polar component as surfactant. When the length of alkyl chain becomes longer, the degree of "dissimilarity" between solvent and surfactant is more significant. On the one hand, increasing the length of hydrophobic chains can lead to a decrease in the water solubility of surfactant monomers, thereby reducing the CMC of surfactant [38]. On the other hand, the surfactant with a longer hydrophobic chain has a higher aggregation number, resulting in larger micelle size and a greater space of the core [39].

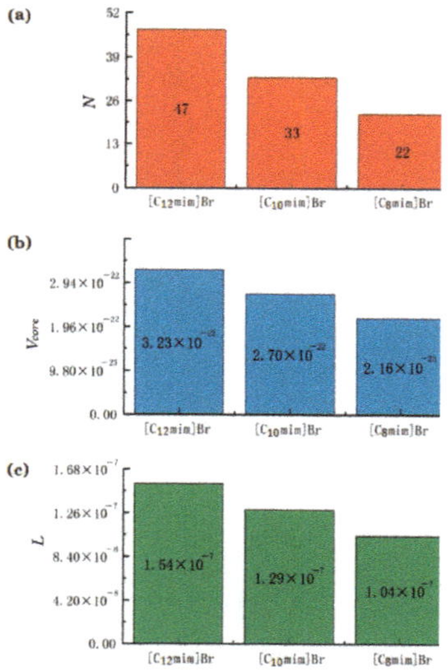

Figure 3. (**a**) The calculated aggregation number (N), (**b**) the volume of single hydrocarbon core (V_{core}), and (**c**) the length of the carbon chain in the micelle core (L) of three imidazolium ILs.

3.2. Effect of Imidazolium ILs on Curcumin Solubility

During the extraction process, using those solvents with a significant solubilization effect can reduce the volume of extractant used and decrease energy consumption and post-treatment burden, thereby improving extraction efficiency. The strengthened solubility of curcumin by ionic liquids is closely related to the aggregation behavior of the ionic liquids themselves. Therefore, the concentration of ionic liquids has a significant impact on the solubility of curcumin, as shown in Figure 4a. When the concentration of $[C_{12}mim]Br$ is below CMC (measured by conductivity method as 10.2 mM), curcumin is almost insoluble. When the concentration exceeds CMC, the solubility significantly increases. This indicates that the solubilization of curcumin by $[C_{12}mim]Br$ is a micelle phenomenon [39].

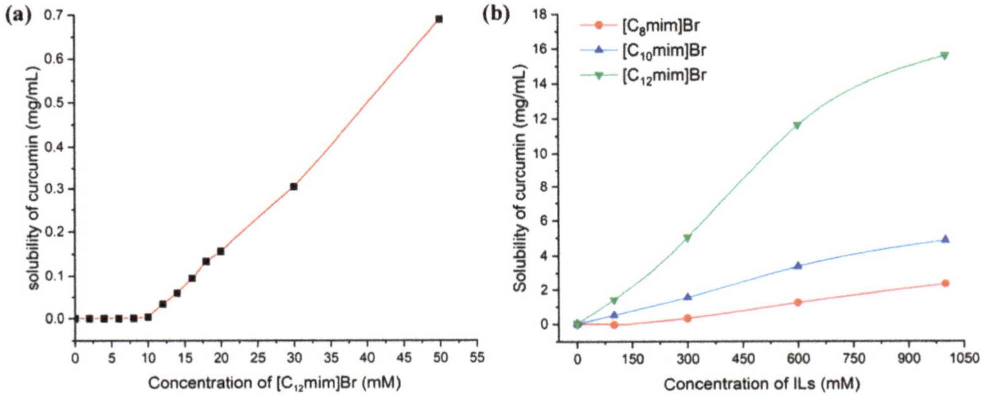

Figure 4. The relationship of curcumin solubility with the concentration of imidazolium ILs (25 °C, (**a**) $[C_{12}mim]Br$ in the range of 0–50 mM, (**b**) three ILs within 1000 mM).

The relationship between the concentration of ionic liquids and the solubility of curcumin in the range of 0–1000 mM is shown in Figure 4b. Through the results, it can be found that with the increase in ionic liquid concentration, the solubilization ability of curcumin significantly increases. This is because as the concentration of ionic liquids increases, on the one hand, the number of micelle aggregates rapidly increases, and more and more curcumin molecules can enter the micelles from the external aqueous phase; on the other hand, the morphology of micelle aggregates began to transform from smaller spherical micelles to larger rod-shaped and layered micelles. The solubilizing space available inside the micelles significantly increased, and the number of soluble curcumin molecules in a single micelle significantly increased. However, when the concentration of the ionic liquid reaches 1000 mM, the water content in the system is already very low, and the viscosity of the system becomes large. The time required for curcumin to reach dissolution equilibrium in the system becomes very long, making the determination of equilibrium solubility more difficult. Figure 5a exhibits the solubility of curcumin in aqueous solutions of $[C_{12}mim]Br$ at different concentrations at 25 °C, 45 °C, and 80 °C. It can be proved that the higher the temperature, the greater the solubility of curcumin. This is due to the increase in temperature leading to an increase in the soluble space in the micelle [28].

On the basis of the above research, considering that ethanol is often used in combination with ionic liquids, we also investigated the solubility of curcumin in imidazole-type ionic liquid ethanol solutions. Due to the good dissolution effect of ethanol on curcumin, using ethanol instead of water to prepare $[C_{12}mim]Br$ ethanolic solution may achieve a better solubilization effect. The relevant experimental results are shown in Figure 5b. At low concentrations, the solubility of curcumin in $[C_{12}mim]Br$ ethanol solution is similar to that in pure ethanol. When the concentration of $[C_{12}mim]Br$ exceeds 50 mM, the solubility begins to significantly increase. Overall, the solubility of curcumin in $[C_{12}mim]Br$ ethanolic solution is higher than that in $[C_{12}mim]Br$ aqueous solution.

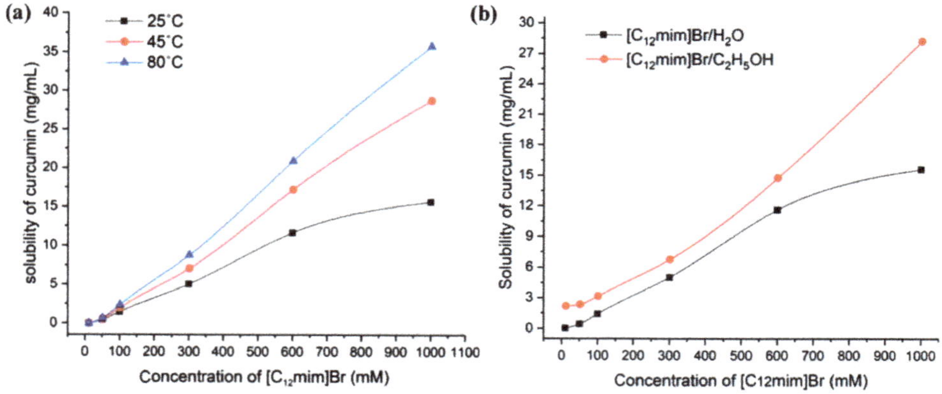

Figure 5. (**a**) The relationship of curcumin solubility with the concentration of $[C_{12}mim]Br$ at different temperatures and (**b**) curcumin solubility in different concentrations of $[C_{12}mim]Br$ aqueous solution and ethanolic solution (25 °C).

3.3. Comparison with the Effect of Tropine-Based ILs on Curcumin Solubility

The solubilization mechanism of Tropinol-based ionic liquids on curcumin is more complex, and it is not only a single effect of micelle solubilization. The aqueous solution of tropine-based ionic liquids has a certain alkalinity, especially when the carbon chain length is less than 10 carbon atoms, its alkalinity is strong [39]. Under strong alkaline conditions, the phenolic hydroxyl groups of curcumin molecules undergo ionization, resulting in a significant increase in molecular polarity and water solubility. Therefore, for Tropine-based ionic liquids with a carbon chain length of 12, the solubilization mechanism is still mainly micelle solubilization, and the solubilization amount of curcumin basically increases linearly with the concentration of the ionic liquid. For Tropine-based ionic liquids with carbon chain lengths of 10 and 8, an increase in their concentration leads to an increase in the alkalinity of the solution, an increase in the ionization degree of curcumin, and, therefore, an increase in solubilization. From Figure 6a, it can be seen that at some concentrations, ionic liquids with short carbon chains actually have greater solubilization. However, under alkaline conditions, curcumin undergoes rapid degradation so the application of such solubilization may be limited. Generally speaking, when curcumin is dissolved in organic solvents, the solution turns yellow. However, in environments with high pH values, the phenolic hydroxyl groups of curcumin undergo ionization, and the solution changes from yellow to red. As shown in Figure 6b, curcumin exhibits different colors when dissolved in different types of ionic liquid aqueous solutions (with curcumin concentrations of 0.5 $mg{\cdot}mL^{-1}$ and ionic liquid concentrations of 200 mM). In the aqueous solutions of $[C_8Tr]Br$ and $[C_{10}Tr]Br$ (sample 2 and 3), the curcumin solution shows a dark red color, indicating that these two ionic liquid aqueous solutions have strong alkalinity. In the aqueous solution of $[C_{12}Tr]Br$ (sample 4), the curcumin solution appears brownish red, indicating that this ionic liquid aqueous solution has weak alkalinity. In ethanol (sample 1) and $[C_{12}mim]Br$ aqueous solutions (sample 5), the curcumin solution appears yellow, and the solution is neutral.

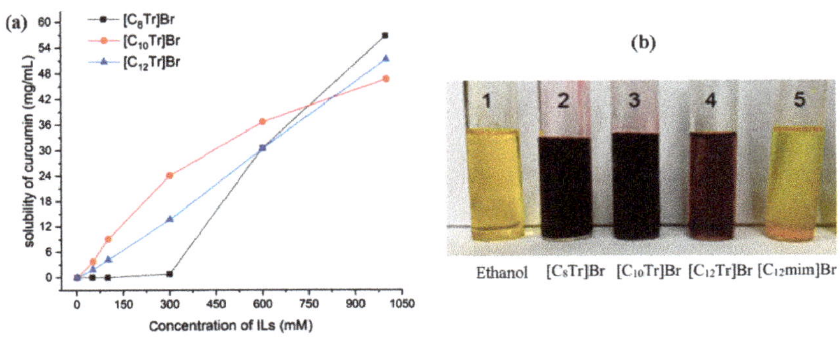

Figure 6. (**a**) The relationship of curcumin solubility with the concentration of three tropine-based ILs (25 °C) and (**b**) color comparison of curcumin solubilized in different solvents.

3.4. Computation for Intermolecular Interactions

In order to further understand the intermolecular interactions between the tested ILs and Cur, Gaussian09W was used to optimize the structures of curcumin and three ionic liquids—[C_8mim]Br, [C_{10}mim]Br, and [C_{12}mim]Br. Initially, the molecular structures of curcumin and the ionic liquid cation were constructed using GaussianView. Subsequently, the DFT-B3LYP method with the 6-31G(d,p) basis set was applied. The system charge was set to +1 to execute the structural optimization. Analyses of weak interactions were performed using the IGMH method implemented in the Multiwfn software (version 3.8) [40,41]. This quantum chemistry-based approach employs the second-largest eigenvalue (λ_2) of the electronic density Hessian matrix, projecting it onto wavefunction eigenvalues that characterize interactions between atoms, among fragments, and within fragments. This procedure generates a color-coded interaction trend map, as illustrated in Figure 7a [42]. The resulting wavefunction data were visualized using VMD software (version 1.9.4a53), producing the depiction shown in Figure 7b. The analysis preliminarily indicates that van der Waals forces are the primary interactions between curcumin and the three ionic liquids, while hydrogen bonds are relatively weaker or nonexistent.

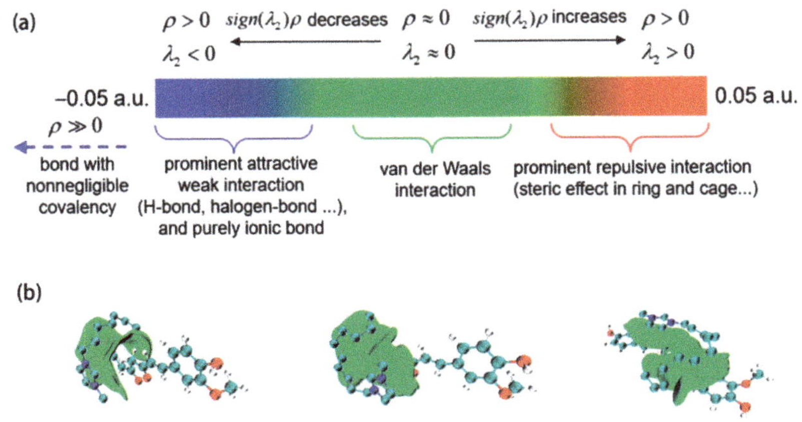

Figure 7. (**a**) IGMH method colorimetric relation; (**b**) VMD visualization IGMH method after the weak interaction analysis results (from left to right: curcumin and C_8mim^+, $C_{10}mim^+$, and $C_{12}mim^+$).

3.5. Electrostatic Potential Distribution of Ionic Liquids and Curcumin Molecules

Molecular electrostatic potential (ESP, V) reflects the overall charge distribution of the system, including both nuclear and electronic charges. It is a true physical property that can be measured experimentally or calculated. ESP is of great significance in investigating intermolecular electrostatic interactions, predicting reaction sites, and predicting molecular properties. By analyzing the ESP distribution on the van der Waals surface, the strength and orientation of many non-covalent interactions can be well predicted and explained, such as hydrogen bonding, halogen bonding, and π-hole bonding [43]. Table 2 exhibits the results from the General Interaction Property Function (GIPF) based on ESP to quantitatively analyze the ESP distribution of ionic liquids and curcumin molecule, respectively, including the highest electrostatic potential ($V_{S,\,max}$), the lowest electrostatic potential ($V_{S,\,min}$), and the average deviation of the distribution of the molecular van der Waals surface (Π) and the variance of the distribution (σ^2). It should be noted that curcumin has two isomers: a diketone structure and an enol structure. The data in Table 2 indicate that the length of the carbon chain of ionic liquids has little effect on the maximum and minimum values of ESP but has a certain impact on the overall distribution of electrostatic potential: (1) For the same type of ionic liquids, the longer the hydrophobic carbon chain is, the more ESP distribution on the van der Waals surface Π. The smaller the carbon chain, the smaller the polarity of ionic liquids with longer carbon chains. (2) When the side chain carbon atoms of the tested imidazole-type ionic liquids exceed 8, σ^2 of the ESP distribution significantly increases, indicating a stronger ability to interact with other molecules through its own charge region. (3) The cationic head group of ionic liquids has the maximum positive charge, while the enol structure of curcumin has the maximum negative charge, and electrostatic interactions may occur between the two. This electrostatic effect may cause damage to the intramolecular hydrogen bonds in the enol structure, causing curcumin molecules to transform from enol-like structures to diketone-like structures.

Table 2. GIPF parameters of ESP distribution on the van der Waals surface of ILs and curcumin.

Objects	$V_{S,\,max}$ (kcal·mol^{-1})	$V_{S,\,min}$ (kcal·mol^{-1})	Π (kcal·mol^{-1})	σ^2 (kcal·mol^{-1})^2
[C_4mim]Br	62.90104	−81.66787	33.40844	767.62555
[C_6mim]Br	63.05364	−81.58291	29.07002	771.87480
[C_8mim]Br	62.33606	−81.63857	26.01044	1225.22731
[C_{10}mim]Br	63.08433	−81.56138	23.44796	1102.42871
[C_{12}mim]Br	63.08845	−81.53636	21.38200	1230.40456
[C_8Tr]Br	59.53544	−83.15667	27.53352	1191.78945
[C_{10}Tr]Br	59.42606	−83.12667	25.01963	1174.60610
[C_{12}Tr]Br	59.38706	−83.10465	22.85953	1153.49930
Enol	47.42085	−47.85527	15.62625	230.34114
Diketone	47.29646	−42.70744	13.48562	202.75778

3.6. Effects of Main Extraction Conditions

Solubilizing enhancement makes some compounds that are originally difficult to extract easier to extract. In addition, it can ensure that the extracted compounds maintain high solubility in the solvent, thereby improving the quality and stability of the product. Strengthened solubilization can also affect the mass transfer rate and equilibrium state during the extraction process, which is of great significance for basic research. In order to achieve rapid extraction, the solvent's ability to dissolve the extract and the rapid mass transfer ability of the substance in the solvent are very important. On the one hand, the solubility of curcumin increases with the increase in ionic liquid concentration. If the concentration of ionic liquid in the extraction agent is too low, the extraction effect will obviously not be good. On the other hand, the viscosity of the extraction agent will also increase with the increase in ionic liquid concentration, and a too-high viscosity will hinder mass transfer and be unfavorable for extraction. Choosing the appropriate conditions is crucial for the extraction of turmeric-like substances.

3.6.1. Effects of Particle Size and Solid-to-Liquid Ratio

Firstly, an ultrasound-assisted extraction of curcuminoids was carried out using ionic liquids with raw materials with different particle sizes (40–120 mesh) under the same conditions. The ionic liquid concentration was 300 mM, the solid–liquid ratio was 1:50 (g·mL^{-1}), the extraction temperature was 30 °C, and the ultrasonic extraction time was 30 min at 100 W. The results are shown in Figure 8a. It can be seen that the total changing trend is to rise first and then gradually reach a level. Usually, a sufficient crushing of raw material particles is necessary, especially for objects such as turmeric that are not easily crushed as small particles are conducive to mass transfer. During crushing, plant cells undergo varying degrees of damage under the action of shear force. Generally, the larger the mesh size, the more obvious the degree of damage, and, at this point, the components within the cell are more likely to detach from the raw material and dissolve in the extraction solvent. However, if crushed too fine, the particles tend to aggregate and float on the surface of the extraction solution without being fully soaked. On the other hand, there are more impurities that can be dissolved. Based on the results of this study, a particle size of 100 mesh is suitable for a sufficient extraction of raw materials.

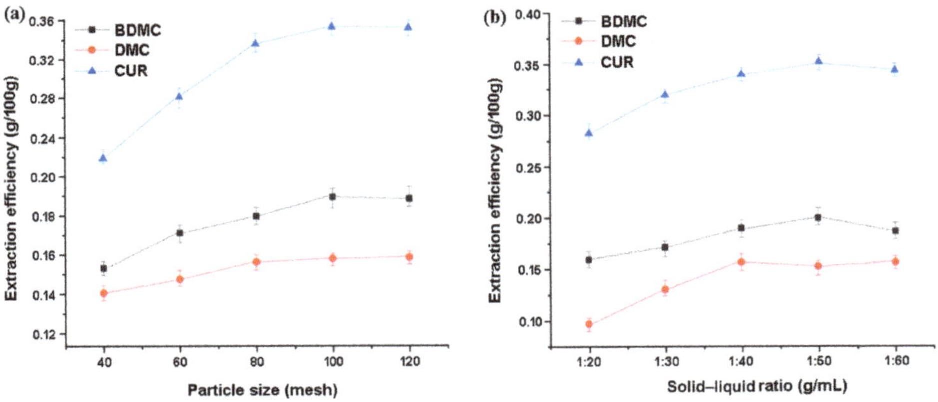

Figure 8. (**a**) The effects of the raw material size and (**b**) solid−liquid ratio on the extraction of curcuminoids.

Moreover, Figure 8b shows the comparison of the different solid–liquid ratios on the target extraction under the conditions including 100 mesh of raw material, 300 mM IL concentration, 30 °C of the extraction temperature, and the 100 W ultrasonic extraction for 30 min. On the one hand, insufficient solvent is hard to achieve thorough extraction; and the increased volume of extractant will result in a greater concentration difference between the solid and liquid phases, which strengthens the driving force of mass transfer in turn. On the other hand, excessive extractant would result in unnecessary wastage and more consumption in post-treatment after extraction; at the same time, voluminous extraction solvents will absorb too much ultrasonic energy, which is not conducive to improving the extraction efficiency. The experimental results indicate that all three components reach high levels at a solid–liquid ratio of 1:50; and the improvement extent of DMC and CUR is more obvious than that of BDMC.

3.6.2. Effects of IL Concentration and Ultrasonic Power

To achieve rapid extraction, the solvent's ability to dissolve the extract and the rapid mass transfer ability of the target substance in the solvent are very important. On the one hand, the solubility of curcumin increases with the increase in ionic liquid concentration. If the concentration of the ionic liquid in the extraction agent is too low, the extraction effect will obviously not be good. On the other hand, the viscosity of the extraction agent will also increase with the increase in ionic liquid concentration, and a too-high viscosity will hinder mass transfer and be unfavorable for extraction. Choosing the appropriate concentration of ionic liquids is crucial for the extraction of curcuminoid substances. Under the same conditions, different concentrations (100–500 mM) of ionic liquids were used as extractants for extraction comparison. In the investigation on IL concentration, the extraction temperature was 30 °C, and the 100 W ultrasonic extraction time was 30 min. Figure 9a shows the trend of the extraction efficiency of curcuminoids with the concentration of ionic liquid [C_{12}mim]Br. It can be proved that 400 mM of [C_{12}mim]Br is enough to achieve the satisfied extraction for the targets.

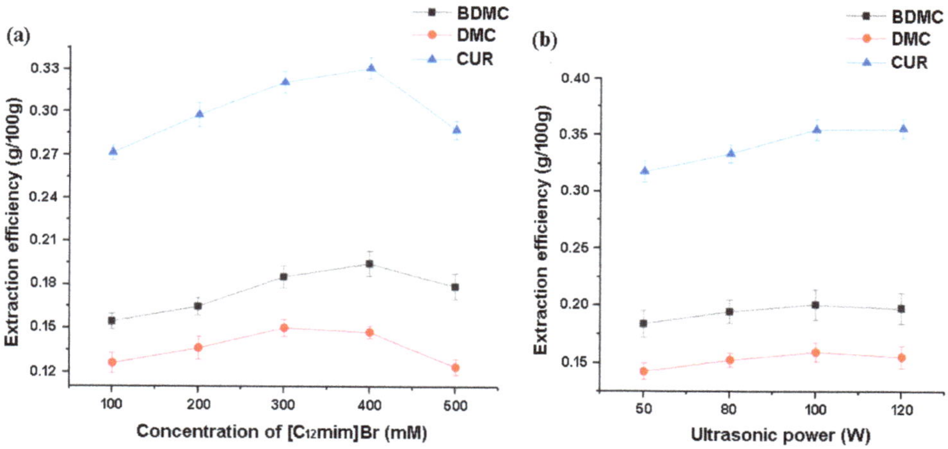

Figure 9. The effects of (**a**) the concentration of [C_{12}mim]Br and (**b**) ultrasonic power on the extraction of curcuminoids.

Ionic liquid ultrasonic extraction method is an extraction method that utilizes the cavitation, mechanical, and thermal effects of ultrasound to accelerate the release, diffusion, and dissolution of intracellular effective substances in ionic liquids, significantly improving extraction efficiency; the cavitation effect is the main driving force for the ultrasonic extraction of ionic liquids. There are often some small bubbles in liquids. When a certain frequency of ultrasound acts on the liquid, small bubbles of appropriate size can resonate. They quickly expand during the sparse stage of ultrasonic wave and are suddenly adiabatic compressed and burst during the compression stage of sound waves. During the process of bursting, small bubbles can generate high-temperature and high-pressure shock waves, which can cooperate with the action of ionic liquids to cause the rupture of materials and even biological cell walls, thereby accelerating the dissolution of intracellular substances. As shown in Figure 9b, an appropriate increase in ultrasound power is beneficial for improving the extraction efficiency, but when the target substances have been fully extracted, excessive power can lead to more coexisting impurities being dissolved. In this study, 100 W is sufficient for the extraction of curcuminoids compounds, which is much lower than the commonly used 200–500 W [44,45] in existing ultrasonic extraction studies by using ionic liquid. This also proves that the selected ionic liquid has a very obvious solubilization effect on the targets, and further increasing the power can only lead to unnecessary energy consumption.

3.7. Study on the Extraction Kinetics of Ionic Liquids

In order to further investigate the impact of extractants on extraction kinetics, different types of ionic liquids and ethanol were used for extraction kinetics studies under the same conditions. The concentration of ionic liquid is 300 mM, and the extraction temperature is 30 °C. During the extraction process, samples are taken every 4 min. The process of extracting turmeric-like substances using ionic liquids can be assumed as two consecutive stages: (1) under the action of ionic liquids, a portion of turmeric-like substances on or near the surface of turmeric powder quickly diffuse and dissolve in the extraction agent; (2) under the dual effects of ionic liquids and ultrasound assistance, more turmeric-like substances are extracted from the inside of turmeric powders. The external diffusion of objects is even slower during this stage [33]. The trend of the concentration of total curcumin substances in the extraction solution (total concentration of curcumin, demethoxycurcumin, and dideoxycurcumin) over time is shown in Figure 10. Before starting ultrasonic extraction (extraction time 0), a certain concentration of turmeric-like substances had already been dissolved in the solution, which corresponds to the first stage mentioned above. In the initial stage of ionic liquid extraction, the extraction rate is the fastest, gradually decreasing over time, and the concentration of turmeric substances eventually approaches saturation.

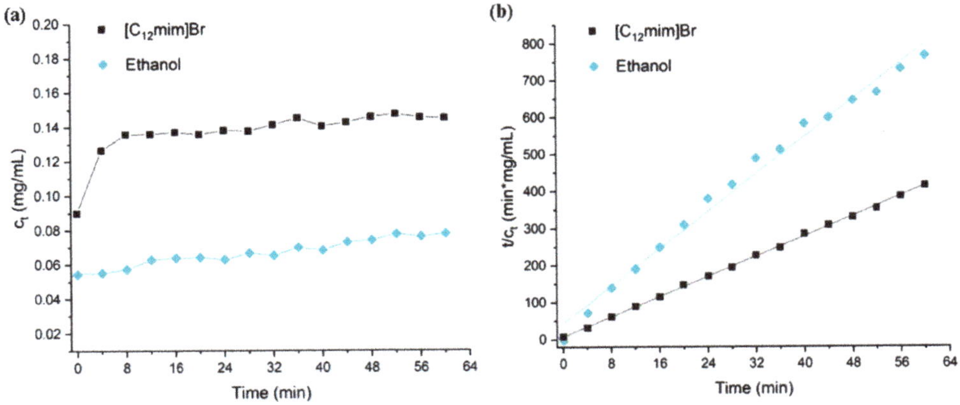

Figure 10. (**a**) Curcuminoid concentration as a function of time in the extraction process with different solvents and (**b**) second-order kinetic fitting for the curcuminoid extraction with different solvents.

From Figure 10a, it can be seen that the ultrasound-assisted extraction of turmeric-like substances is very important. In the process of extracting turmeric substances using ionic liquids, with the increase in ultrasound time, the concentration of turmeric substances in the extraction solution initially increases rapidly and finally tends to flatten out. Without ultrasound assistance, the extraction efficiency of ionic liquids will significantly decrease. When the extraction time is 0, the concentration of turmeric substances dissolved in [C_{12}mim] Br is higher than that in ethanol. The first stage is related to the type of ionic liquid but not to the length of the carbon chain. It may be due to the stronger hydrogen bonding effect of imidazolium ionic liquids compared to quaternary ammonium ionic liquids, which have a stronger ability to dissolve and destroy cellulose in the cell wall. The second stage is not only related to the type of ionic liquid but also to the carbon chain length of the ionic liquid as the mass transfer process is dominant in this stage. Increasing the carbon chain length of the ionic liquid leads to an increase in viscosity, and slowing down the mass transfer leads to a slower extraction rate. In the process of ethanol extraction, the concentration of turmeric substances shows a relatively gentle trend over time, and the final extraction effect is not as good as that of ionic liquids. Due to the poor penetration of ethanol compared to ionic liquids, the extraction ability of turmeric-like substances in turmeric powder is poor, and the extraction efficiency in the second stage is poor. Simulate the second stage of the extraction process using a second-order kinetic model of extraction. The simulated linear fitting results are shown in Figure 10b. Moreover, the extraction saturation concentration (C_S), extraction rate constant (k), initial extraction rate (h), and correlation coefficient (R^2) obtained through fitting calculation are shown in Table 3. It can be observed that the initial extraction rate and rate constant of ionic liquids are higher than those of ethanol. As comparison, the R^2 of first-order kinetics was 0.89861, indicating the latter was not suitable for the fitting data here.

Table 3. Second-order kinetic fitting data of the curcuminoid extraction with different solvents.

Extractant	Slop	Intercept	C_s (mg·mL^{-1})	h (mg·mL^{-1}min^{-1})	k (mL·mg^{-1}min^{-1})	R^2
[C_{12}mim]Br	6.79958	6.68604	0.14707	0.14957	6.91505	0.99893
Ethanol	12.60328	44.53841	0.07934	0.02245	3.56642	0.98852

3.8. ComplexGAPI Analysis and Product Antioxidant Activity

A prominent approach in chemistry is the philosophy of green chemistry, which aims to conduct chemical processes in accordance with the principles of sustainable development. Greenness is a crucial parameter for evaluating experimental methods. The degree of greenness can be assessed through various tools, such as Life Cycle Assessment (LCA) and the Green Analytical Procedure Index (GAPI). The ComplexGAPI is a tool that encompasses all aspects of the analytical process, from sample collection, transportation, preservation, and storage to sample preparation and final analysis, also including those aspects and processes conducted prior to general separation methods [46,47]. The creation of ComplexGAPI is based on the same principles that guided the development of GAPI. The analytical results illustrated in Figure 11a demonstrate the eco-friendliness of the entire extraction and separation process, encompassing sample preparation, reagent and chemical usage, and the collection, preservation, transportation, and storage of samples. It can be intuitively observed that this work aligns with the principles of green chemistry under most conditions. Compared to GAPI, ComplexGAPI features an additional hexagonal region at the base of its diagram, symbolizing the "green" attributes of the pre-analytical process. This region encompasses factors such as yield, conditions, reagents, solvents, instrumentation, and the processing and purification of the product. The tool employs a color scale to evaluate the environmental impact at each stage, ranging from green to yellow to red, with green indicating compliance with predefined numerical standards. For instance, when the yield exceeds 89%, its corresponding position in the diagram is displayed in green; for yields between 70% and 89%, the color appears yellow; and for yields below 70%, it is represented in red. The specific evaluation criteria can be found in Table 1 in [46]. By employing green assessment methods, an experimental design can be utilized to minimize or eliminate the use or generation of hazardous substances, facilitating the discovery of techniques and approaches to accelerate chemical reactions with minimal reagent use while achieving equivalent results at the same cost. Additionally, evaluating the life cycle of chemical products—including their design, manufacturing, usage, and ultimate disposal—can help to reduce synthesis steps, increase yields, and enhance plant capacity, all while lowering energy and water consumption. The use of toxic and/or hazardous reagents and solvents can be avoided by substituting them with low-toxicity or non-toxic alternatives, thereby reducing chemical waste generation. In essence, green chemistry analytical tools enable researchers to swiftly identify shortcomings in the processes under consideration and pinpoint critical areas that require attention to mitigate environmental harm and ensure personal safety. In our work, the experimental design was also guided by the principles of green chemistry. We used a method that operates under green solvent conditions at room temperature, featuring high atom economy and recyclability. This approach effectively minimizes environmental pollution and reduces potential impacts on operators. Furthermore, the extracted products using ethanol and [C_{12}mim]Br were compared with their antioxidant activity based on DPPH method [48] (see the details in Section S3 of Supplementary Materials), and the extract using the latter was obtained by back-extraction with methylene chloride and concentration under vacuum. At the levels of three concentrations, the antioxidant activity of [C_{12}mim]Br-based extract is better than that of ethanol-based extract in the high, medium, and low concentrations below 12 mg·mL^{-1}, which can further prove the significance of IL-enhanced process. In summary, the solubilization effect of ionic liquids has multiple significance and roles in the field of extraction. By the rational utilization of enhanced solubilization, extraction efficiency can be improved and extraction conditions can be optimized, and then application scope can be expanded; as the results, it is beneficial for improving product quality and stability, as well as promoting environmental protection and sustainable development.

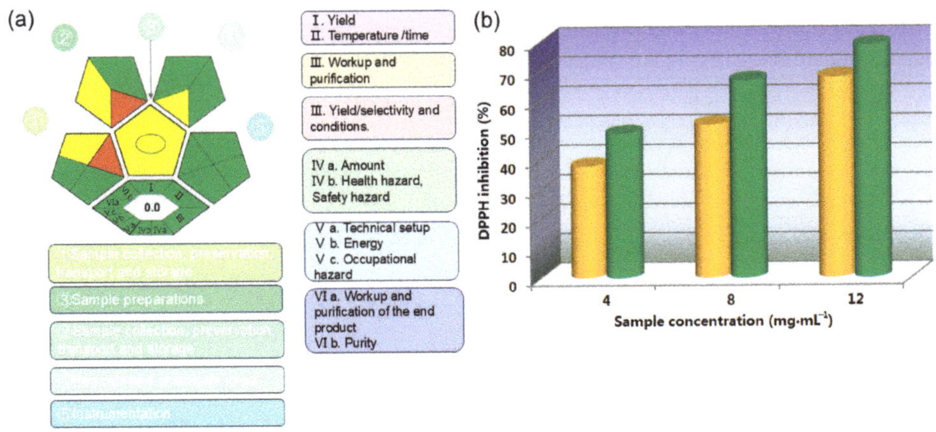

Figure 11. (**a**) The greenness of extraction evaluated by ComplexGAPI and (**b**) antioxidant activities of different extracts (yellow: ethanol extract; green: IL extract).

4. Conclusions

In this study, we determined the critical micelle concentration of imidazolium ionic liquids in water using a conductivity method and calculated the micelle aggregation number using a geometric method. Secondly, the solubility of curcumin in ionic liquid aqueous solution was determined, and the structure of the ionic liquid and the effect of ionic liquid concentration on the solubility of curcumin were explored. The results indicate that the solubilization ability of ionic liquids for curcumin is closely related to their micelle aggregation behavior. The longer the carbon chain and higher the concentration of ionic liquids, the higher the solubility of curcumin. In addition, the size and charge type of the hydrophilic structure of ionic liquid cations also have a certain impact on the solubility of curcumin. At concentrations below the critical micelle concentration, there is an electrostatic interaction between the monomeric ionic liquid and curcumin, leading to the transformation of curcumin from an enol structure to a diketone structure. After reaching the critical micelle concentration, the ionic liquid aggregates to form micelles. Computational chemistry further reveals the key interactions between the two. In ultrasonic extraction, the extraction rate of turmeric-like substances showed a trend of first increasing and then decreasing with the increase in ionic liquid concentration. The optimal extraction concentrations of [C_{12}mim] Br were 400 mM and 500 mM, respectively. In addition, ideal results were obtained by fitting with the second-order kinetic model, and the rate constant and initial extraction rate during the extraction process were obtained. Finally, ComplexGAPI analysis and product antioxidant activity have also proved the advantage of IL-based extraction.

Supplementary Materials: The following supporting information can be downloaded at https://www.mdpi.com/article/10.3390/separations12020029/s1. Figure S1: 1H NMR profile of [Cnmim]Br (300 MHz, D_2O), Figure S2: 1H NMR profile of [CnTr]Br (300 MHz, D_2O).

Author Contributions: Conceptualization, D.L. and Y.Q.; methodology, D.L.; software, Y.C.; validation, Y.C. and J.L.; formal analysis, J.L.; investigation, D.L.; data curation, Y.Q.; writing—original draft preparation, S.M.; writing—review and editing, S.Y.; visualization, J.S.; supervision, S.Y.; project administration, S.Y. and Y.C.; funding acquisition, S.Y. and Y.C. All authors have read and agreed to the published version of the manuscript.

Funding: This work is supported by Natural Science Foundation of Sichuan Province (2022NSFSC1604) and Fundamental Research Funds for the Central Universities/Sichuan University-Luzhou Science and Technology Innovation Platform Construction Project (2022CDLZ-20).

Data Availability Statement: Data is contained within the article or Supplementary Materials.

Acknowledgments: All the authors' affiliations provided the convenience for related studies, respectively. Special thanks to the Engineering Experimental Teaching Center, School of Chemical Engineering, Sichuan University for related measurements.

Conflicts of Interest: The authors declare no conflicts of interest.

References

1. Perinelli, D.R.; Cespi, M.; Lorusso, N.; Palmieri, G.F.; Bonacucina, G.; Blasi, P. Surfactant Self-Assembling and Critical Micelle Concentration: One Approach Fits All? *Langmuir* **2020**, *36*, 5745–5753. [CrossRef]
2. Williams, H.D.; Trevaskis, N.L.; Charman, S.A.; Shanker, R.M.; Charman, W.N.; Pouton, C.W.; Porter, C.J.H. Strategies to Address Low Drug Solubility in Discovery and Development. *Pharmacol. Rev.* **2013**, *65*, 315–499. [PubMed]
3. Rangel-Yagui, C.O.; Pessoa, A.; Tavares, L.C. Micellar solubilization of drugs. *J. Pharm. Pharm. Sci.* **2005**, *8*, 147–163. [PubMed]
4. Fayyaz, S.; Talat, R.; Ali, S.; Khalid, N.; Shah, A.; Ullah, F. Synthesis, Characterization, and Micellization Behavior of Cationic Surfactants: N-Alkyl-3-Methylpyridinium Bromides and Their Drug Interaction Study by UV–Visible Spectroscopy and Conductometry. *J. Surfactants Deterg.* **2019**, *22*, 625–632. [CrossRef]
5. Jabbari, M.; Teymoori, F. An insight into effect of micelle-forming surfactants on aqueous solubilization and octanol/water partition coefficient of the drugs gemfibrozil and ibuprofen. *J. Mol. Liq.* **2018**, *262*, 1–7. [CrossRef]
6. Vinarov, Z.; Katev, V.; Radeva, D.; Tcholakova, S.; Denkov, N.D. Micellar solubilization of poorly water-soluble drugs: Effect of surfactant and solubilizate molecular structure. *Drug Dev. Ind. Pharm.* **2018**, *44*, 677–686. [CrossRef] [PubMed]
7. Zhuo, Y.; Cheng, H.L.; Zhao, Y.G.; Cui, H.R. Ionic Liquids in Pharmaceutical and Biomedical Applications: A Review. *Pharmaceutics* **2024**, *16*, 151. [CrossRef]

Buettner, C.S.; Cognigni, A.; Schroder, C.; Bica-Schroder, K. Surface-active ionic liquids: A review. *J. Mol. Liq.* **2022**, *347*, 118160. [CrossRef]

Sarac, B.; Medos, Z.; Cognigni, A.; Bica, K.; Chen, L.J.; Bešter-Rogač, M. Thermodynamic study for micellization of imidazolium based surface active ionic liquids in water: Effect of alkyl chain length and anions. *Colloids Surf. A-Physicochem. Eng. Asp.* **2017**, *532*, 609–617. [CrossRef]

10. Kuznetsova, D.A.; Gabdrakhmanov, D.R.; Lukashenko, S.S.; Voloshina, A.D.; Sapunova, A.S.; Kulik, N.V.; Nizameev, I.R.; Kadirov, M.K.; Kashapov, R.R.; Zakharova, L.Y. Supramolecular systems based on cationic imidazole-containing amphiphiles bearing hydroxyethyl fragment: Aggregation properties and functional activity. *J. Mol. Liq.* **2019**, *289*, 111058. [CrossRef]
11. Dong, B.; Zhao, X.Y.; Zheng, L.Q.; Zhang, J.; Li, N.; Inoue, T. Aggregation behavior of long-chain imidazolium ionic liquids in aqueous solution: Micellization and characterization of micelle microenvironment. *Colloids Surf. A-Physicochem. Eng. Asp.* **2008**, *317*, 666–672. [CrossRef]
12. Baltazar, Q.Q.; Chandawalla, J.; Sawyer, K.; Anderson, J.L. Interfacial and micellar properties of imidazolium-based monocationic and dicationic ionic liquids. *Colloids Surf. A-Physicochem. Eng. Asp.* **2007**, *302*, 150–156. [CrossRef]
13. Vanyur, R.; Biczok, L.; Miskolczy, Z. Micelle formation of 1-alkyl-3-methylimidazolium bromide ionic liquids in aqueous solution. *Colloids Surf. A-Physicochem. Eng. Asp.* **2007**, *299*, 256–261. [CrossRef]
14. Inoue, T.; Ebina, H.; Dong, B.; Zheng, L.Q. Electrical conductivity study on micelle formation of long-chainimidazolium ionic liquids in aqueous solution. *J. Colloid Interface Sci.* **2007**, *314*, 236–241. [CrossRef] [PubMed]
15. Dong, B.; Li, N.; Zheng, L.Q.; Li, Y.; Inoue, T. Surface adsorption and micelle formation of surface active ionic liquids in aqueous solution. *Langmuir* **2007**, *23*, 4178–4182. [CrossRef]
16. Zhao, Y.; Gao, S.J.; Wang, J.J.; Tang, J.M. Aggregation of ionic liquids [C(n)mim]Br (n = 4, 6, 8, 10, 12) in D_2O: A NMR study. *J. Phys. Chem. B* **2008**, *112*, 2031–2039. [CrossRef]
17. Sirieix-Plenet, J.; Gaillon, L.; Letellier, P. Behaviour of a binary solvent mixture constituted by an amphiphilic ionic liquid, 1-decyl-3-methylimidazolium bromide and water Potentiometric and conductimetric studies. *Talanta* **2004**, *63*, 979–986. [CrossRef]
18. Goodchild, I.; Collier, L.; Millar, S.L.; Prokeš, I.; Lord, J.C.D.; Butts, C.P.; Bowers, J.; Webster, J.R.P.; Heenan, R.K. Structural studies of the phase, aggregation and surface behaviour of 1-alkyl-3-methylimidazolium halide plus water mixtures. *J. Colloid Interface Sci.* **2007**, *307*, 455–468. [CrossRef]
19. Mester, P.; Wagner, M.; Rossmanith, P. Ionic liquids designed as chaotrope and surfactant for use in protein chemistry. *Sep. Purif. Technol.* **2012**, *97*, 211–215. [CrossRef]
20. Alawi, M.A.; Hamdan, I.I.; Sallam, A.A.; Heshmeh, N.A. Solubility enhancement of glibenclamide in choline–tryptophan ionic liquid: Preparation, characterization and mechanism of solubilization. *J. Mol. Liq.* **2015**, *212*, 629–634. [CrossRef]
21. Pathan, H.; Patil, R.; Ray, D.; Aswal, V.K.; Bahadur, P.; Tiwari, S. Structural changes in non-ionic surfactant micelles induced by ionic liquids and application thereof for improved solubilization of quercetin. *J. Mol. Liq.* **2019**, *290*, 111235. [CrossRef]
22. Pino, V.; German-Hernandez, M.; Martin-Perez, A.; Anderson, J.L. Ionic Liquid-Based Surfactants in Separation Science. *Sep. Sci. Technol.* **2012**, *47*, 264–276. [CrossRef]
23. Amalraj, A.; Pius, A.; Gopi, S.; Gopi, S. Biological activities of curcuminoids, other biomolecules from turmeric and their derivatives—A review. *J. Tradit. Complement. Med.* **2016**, *15*, 205–233. [CrossRef] [PubMed]
24. Shahrajabian, M.H.; Sun, W. The Golden Spice for Life: Turmeric with the Pharmacological Benefits of Curcuminoids Components, Including Curcumin, Bisdemethoxycurcumin, and Demethoxycurcumins. *Curr. Org. Synth.* **2024**, *21*, 665–683. [CrossRef]
25. Wu, H.; Liu, Z.; Zhang, Y.; Gao, B.; Li, Y.; He, X.; Sun, J.; Choe, U.; Chen, P.; Blaustein, R.A.; et al. Chemical Composition of Turmeric (*Curcuma longa* L. *) Ethanol Extract and Its Antimicrobial Activities and Free Radical Scavenging Capacities. Foods* **2024**, *13*, 1550.
26. Cornellas, A.; Perez, L.; Comelles, F.; Ribosa, I.; Manresa, A.; Garcia, M.T. Self-aggregation and antimicrobial activity of imidazolium and pyridinium based ionic liquids in aqueous solution. *J. Colloid Interface Sci.* **2011**, *355*, 164. [CrossRef] [PubMed]
27. Paterno, A.; Fiorenza, R.; Marullo, S.; Musumarra, G.; Scire, S. Prediction of ionic liquid's heat capacity by means of their in silico principal properties. *RSC Adv.* **2016**, *6*, 36085. [CrossRef]
28. Lu, J.; Song, H.; Yang, Y.; Qian, G.F.; Nie, L.R.; Yao, S. Synthesis and characterization of novel hexafluorophosphate salts with tropine-type cations. *J. Mol. Liq.* **2015**, *209*, 648–656. [CrossRef]
29. Pal, A.; Chaudhary, S. Ionic liquids effect on critical micelle concentration of SDS: Conductivity, fluorescence and NMR studies. *Fluid Phase Equilibria* **2014**, *327*, 100–104. [CrossRef]
30. Rahman, M. Development of UV spectrophotometric method for estimation of curcumin in bulk drug and nanogel formulation: A hydrolytic degradation studies. *Int. J. Pharm. Chem. Anal.* **2022**, *9*, 87–92. [CrossRef]
31. Xu, J.; Wang, W.; Liang, H.; Zhang, Q.; Li, Q.Y. Optimization of ionic liquid based ultrasonic assisted extraction of antioxidant compounds from Curcuma longa L. using response surface methodology. *Ind. Crops Prod.* **2015**, *76*, 487–493. [CrossRef]

32. Liang, H.; Wang, W.; Xu, J.; Zhang, Q.; Shen, Z.; Zeng, Z.; Li, Q. Optimization of ionic liquid-based microwave-assisted extracti technique for curcuminoids from *Curcuma longa* L. *Food Bioprod. Process.* **2017**, *104*, 57–65. [CrossRef]
33. Bogdanov, M.G.; Svinyarov, I. Ionic liquid-supported solid–liquid extraction of bioactive alkaloids. II. Kinetics, modeling a mechanism of glaucine extraction from *Glaucium flavum* Cr. (Papaveraceae). *Sep. Purif. Technol.* **2013**, *103*, 279–288. [CrossRef
34. Ali, A.; Lim, X.Y.; Chong, C.H.; Mah, S.H.; Chua, B.L. Ultrasound-assisted extraction of natural antioxidants from betel leav (*Piper betle*): Extraction kinetics and modeling. *Sep. Sci. Technol.* **2018**, *53*, 2192–2205. [CrossRef]
35. Gonzalez-Perez, A.; Del Castillo, J.L.; Czapkiewicz, J.; Rodriguez, J.R. Conductivity, density, and adiabatic compressibility dodecyldimethylbenzylammonium chloride in aqueous solutions. *J. Phys. Chem. B* **2001**, *105*, 1720–1724. [CrossRef]
36. Tanford, C. Micelle Shape and Size. *J. Phys. Chem.* **1972**, *76*, 3020–3024. [CrossRef]
37. Wang, J.J.; Wang, H.Y.; Zhang, S.L.; Zhang, H.C.; Zhao, Y. Conductivities, volumes, fluorescence, and aggregation behavior ionic liquids [C(4)mim][BF4] and [C(n)mim]Br(n = 4, 6, 8, 10, 12) in aqueous solutions. *J. Phys. Chem. B* **2007**, *111*, 6181–618 [CrossRef] [PubMed]
38. Chandler, D. Interfaces and the driving force of hydrophobic assembly. *Nature* **2005**, *437*, 640–647. [CrossRef]
39. Rosen, M.J.; Kunjappu, J.T. *Surfactants and Interfacial Phenomena*; Wiley: Hoboken, NJ, USA, 2012.
40. Lu, T.; Chen, F.W. Multiwfn: A Multifunctional Wavefunction Analyzer. *J. Comput. Chem* **2012**, *33*, 580–592. [CrossRef]
41. Lu, T. A comprehensive electron wavefunction analysis toolbox for chemists, Multiwfn. *J. Chem. Phys.* **2024**, *161*, 0825C [CrossRef]
42. Lu, T.; Chen, Q.X. Independent gradient model based on Hirshfeld partition: A new method for visual study of interactions chemical systems. *J. Comput. Chem.* **2022**, *43*, 539–555. [CrossRef] [PubMed]
43. Zhang, Y.F.; Zhou, Z.L.; Zou, L.; Chi, R. Imidazolium-based ionic liquids with inorganic anions in the extraction of salidrosid and tyrosol from Rhodiola: The role of cations and anions on the extraction mechanism. *J. Mol. Liq.* **2019**, *275*, 136–145. [CrossRe
44. Sui, X.Y.; Liu, T.T.; Liu, J.C.; Zhang, J.; Zhang, H.L.; Wang, H.Y.; Yang, Y. Ultrasonic-enhanced surface-active ionic liquid-base extraction and defoaming for the extraction of psoralen and isopsoralen from *Psoralea corylifolia* seeds. *Ultrason. Sonochem.* **202** *69*, 105263. [CrossRef]
45. Tan, Z.-J.; Wang, C.-Y.; Yang, Z.-Z.; Yi, Y.-J.; Wang, H.-Y.; Zhou, W.-L.; Li, F.-F. Ionic Liquid-Based Ultrasonic-Assisted Extractio of Secoisolariciresinol Diglucoside from Flaxseed (*Linum usitatissimum* L.) with Further Purification by an Aqueous Two-Phas System. *Molecules* **2015**, *20*, 17929–17943. [CrossRef] [PubMed]
46. Płotka-Wasylka, J.; Wojnowski, W. Complementary green analytical procedure index (ComplexGAPI) and software. *Green Chem* **2021**, *23*, 8657–8665. [CrossRef]
47. El-Eryan, R.T.; Toubar, S.S.; Ashour, A.A.; Elshahed, M.S. Application of analytical Eco-Scale and Complex-GAPI tools for gree assessment of a new simple nanoparticle modified carbon paste electrode method for voltammetric determination of mosaprid citrate in pharmaceutical dosage form and human plasma. *Microchem. J.* **2022**, *178*, 107347. [CrossRef]
48. Ak, T.; Gülçin, I. Antioxidant and radical scavenging properties of curcumin. *Chem. Biol. Interact.* **2008**, *174*, 27–37. [CrossRef]

separations

Article

Process Design and Techno-Economic Analysis of Heat Pump-Assisted Distillation for Crude Phenol Separation

Dechang Meng [1,2,3,†], Liying Qin [1,2,3,†], Yuan Zhao [1,2,3], Jiawei Zhao [1,2,3], Chunping Yan [1,2,3], Chenghong Mou [4], Jieming Xiong [4,*] and Chen Zhang [4,*]

1 CCTEG China Coal Research Institute, Chaoyang, Beijing 100013, China
2 National Energy Key Laboratory of Coal Utilization and Emission Control Technology and Equipment, Chaoyang, Beijing 100013, China
3 CCTEG Low-Carbon Technology Research Institute, Chaoyang, Beijing 100013, China
4 Beijing Key Laboratory of Enze Biomass Fine Chemicals, College of New Materials and Chemical Engineering, Beijing Institute of Petrochemical Technology, Beijing 102617, China
* Correspondence: xjm@bipt.edu.cn (J.X.); zhangc@bipt.edu.cn (C.Z.)
† These authors contributed equally to this work.

Abstract: In China, crude phenols, mixtures commonly produced in the coal industry, are inexpensive and abundant in supply, but their valorization is hindered by high energy consumption in the separation process. It is of great academic and commercial significance to improve the separation process of crude phenols to achieve energy efficiency and cost reduction. In this study, a heat pump-assisted distillation (HPD) system for crude phenol separation was developed. External vapor recompression was adopted due to the strong corrosiveness, high toxicity, heat sensitivity, and easy polymerization of crude phenols. Compared with conventional distillation (CD), HPD showed clear advantages in lowering operating costs. The effects of design variables including pressure, the number of theoretical plates and temperature differences between the condenser and reboiler on reflux ratios, kettle temperature, equipment costs, operating costs, and total annual cost (TAC) were investigated and optimized in detail. The effect of steam prices on process economic feasibility was also studied. It was found that HPD reduced at least 55% of the operational cost compared to CD when the steam price was higher than 10.8 USD/GJ. Carbon emission evaluation indicated that CO_2 generated by the HPD process was 56.3% lower than CD.

Keywords: crude phenols; heat pump assisted distillation; techno-economic analysis

1. Introduction

Crude phenol is a mixture primarily composed of phenol and its homologues, which are mainly presented in coal tar from many coal chemical processes and in wastewater from semicoke production. Crude phenols can be extracted from coal tar by alkali washing [1–3] or from wastewater using organic solvent extraction [4–7]. After extraction, the general strategy for crude phenol purification is vacuum distillation with a high column to yield pure compound or isomeric mixtures, such as phenol, cresol or xylenol [8]. These compounds are important raw materials to produce bisphenol A [9], bisphenol F [10], and phenolic epoxy resin [11], etc. However, due to the small difference in boiling points among components in crude phenols and thus the great difficulty in separation, the distillation of crude phenol is an energy-intensive process with large reflux ratios and high steam consumption, resulting in high production costs and low economic value [8]. Processes with reduced energy consumption and operation costs attract the attention of both academia and industry.

Heat pump-assisted distillation (HPD) as an energy efficient distillation technology has shown promising potential in recent years. Several articles have thoroughly reviewed recent progress [12–14], and the advantages and disadvantages of HPD are briefly summarized in Table 1. By converting low-quality heat from the condenser to high-quality heat via vapor compression, HPD can utilize the heat released from the condenser for heating in the reboiler at the expense of a relatively small amount of electricity in the compressor. For crude phenol separation, where a large amount of heat is required for reboilers, HPD could greatly reduce steam consumption. Therefore, researchers have attempted to use HPD for crude phenol separations [8,15]. The most typical example is the research from Gao et al. [8], conducting a comparative analysis of VRC (vapor recompression distillation column), IMVRC (improved middle vapor recompression distillation column), and other crude phenol separation technologies combined with heat pumps. It was discovered that the utilization of IMVRC in crude phenol processing managed to reduce the total annual cost (TAC) by 41.76–74.92% depending on different payback years [8].

Table 1. The advantages and disadvantages of heat pump-assisted distillation compared to conventional distillation.

Advantages	Disadvantages
Reducing energy consumption by recovering heat	High capital investment and longer payback periods
Lower operation costs	Sensitive to energy prices
Reduce CO_2 emission and beneficial for environment	Difficulty in process integration

Although HPD is undoubtedly energy-saving for crude phenol separation, some severe issues in crude phenol processing are still unsolved. Crude phenols are corrosive, heat-sensitive, and chemically unstable under high temperatures. The industrial distillation of crude phenols often results in asphalt-like phenol residues, which may cause blockages in equipment, pipelines, valves, and instruments. Such characteristics of crude phenols greatly increase the difficulty of distillation design and operation, particularly for HPD using high-temperature vapor. For instance, due to the corrosion and instability of phenolic compounds, it is impractical to compress phenolic vapor or handle phenols at high temperature (typically above 180 °C), while previous literature references did not consider these constraints. The separation of phenolic compounds is also rather challenging, which requires numerous industrial experiences. Therefore, the refining process of crude phenols in combination with the heat pump-assisted distillation should be carried out and optimized in consideration of these features of phenolic compounds, to propose feasible separation strategy, to achieve practical improvement in separation efficiency, and to reduce energy consumption. Another disadvantage of HPD is the high capital costs for compressors. Compared to other equipment, such as columns or heat exchangers, steam compressors are quite expensive, which greatly increases the payback period of the HPD process despite the reduced operation costs. Compressors working with steam at relatively high temperatures may cost a million dollars [16]. For example, HPD might cost more than conventional distillation (CD) in regions where steam is inexpensive and readily obtainable, so that TAC of HPD process becomes higher than CD due to elevated equipment capital costs.

There have been a few in-depth studies on crude phenol separation in the past, as well as a lack of analysis on the effect of crucial parameters such as column pressure or theoretical stages on TAC. Moreover, the application of heat pumps in crude phenol separation in previous research is not only scarce but also fails to consider practical issues that cannot be ignored during the crude phenol separation process, such as thermal sensitivity, coking, and corrosion. Therefore, it is of great significance to explore the application of HPD technology on crude phenol separation based on the existing production schemes, to achieve reduced energy consumption, and to lower TAC. In this work, separation schemes using CD and HPD were designed and analyzed to achieve maximum productivity using crude phenol from coal tar as the feed. The capital cost and operation cost of each process were carefully estimated to ensure authenticity. A comprehensive techno-economic evaluation was conducted to demonstrate the advantages and disadvantages of each process. The effect of operation parameters on separation and cost was investigated.

2. Methodology

2.1. Process Design and Simulation

Crude phenols used in this study were from coal tar in Naomaohu, Xinjiang, China. Typically, phenol oil fractions from coal tar with boiling points of 170–220 °C are counter-currently extracted by a binary solvent composed of a polar solvent and a non-polar solvent to remove neutral oils, followed by distillation to recover solvents and obtain crude phenol with minor impurities (pyridines, anilines, alcohols, ketones, thiophenols, etc.) [17]. These impurities were eliminated using chemical methods to produce the crude phenol mixture used in this work. Crude phenols after extraction were free of water and solid residue, and the typical composition and boiling points of components are shown in Table 2. The flow rate of crude phenols was 3125 kg/h.

Table 2. Composition and boiling points of components in crude phenols.

Components	Content (wt%)	Boiling Points (°C) [a]
Phenol	12.76	181.8
o-Cresol	12.19	191.0
m-Cresol	18.02	202.2
p-Cresol	16.98	201.9
2,3-Xylenol	1.44	216.9
2,4-Xylenol	8.71	210.9
2,5-Xylenol	5.17	211.1
2,6-Xylenol	5.03	201.0
3,4-Xylenol	0.44	227.3
3,5-Xylenol	7.58	221.7
o-Ethylphenol	3.37	204.5
m-Ethylphenol	7.50	218.4
p-Ethylphenol	0.82	218.0

[a] Adapted from source [18].

The separation processes of crude phenols were simulated by Aspen Plus simulation (V12) software. Considering that crude phenols were polar systems, the non-random two-liquid (NRTL) model was the thermodynamic model for the simulations. Default binary interaction parameters in the software were used. Missing physical property parameters were estimated by the universal functional activity coefficient (UNIFAC) method. The RadFrac module in the software was used to simulate the distillation column in the process [19,20]. All columns were packed with BX gauze packing of 316 L stainless steel. The theoretical plate number of each column was set to 150 for all columns due to the high difficulty in phenol separation. In the simulation, the pressure drop of all condensers was fixed at 3 kPa and the pressure drop of all reboilers was neglected. The pressure difference in the whole column (excluding condensers and reboilers) was 8 kPa unless otherwise specified. These values were mainly based on industrial experience from crude phenol plants. To ensure sufficient heat transfer rate, the temperature difference for condensers and reboilers were set to be 10 and 15 °C, respectively, unless otherwise specified. The inlet temperature of cooling water in this study was 32 °C and the outlet temperature was 40 °C. The heat transfer area of heat exchangers was estimated using Aspen Exchangers Design and Rating (V12). The isentropic efficiency of steam compressors or vacuum pumps was set to be 0.72 to estimate their power and electricity usage (based on the actual isentropic efficiency from realistic compressors).

2.2. Economic Analysis and CO_2 Emission Calculations

The financial performance of conventional distillation and heat pump-assisted distillation was assessed with total capital cost (TCC), total operating cost (TOC), and total annual cost (TAC). TCC was composed of capital costs of installed equipment, including columns, heat exchangers, packing materials, vacuum pumps, and compressors. TOC was contributed by the electricity consumption by vacuum pumps and compressors, steam costs for reboilers, and cooling water cost for condensers. The costs of instrumentation, piping, paintings, electrical and others were not considered, as these costs were complicated to estimate and also did not affect the conclusions made in this study. The basis and factors used for calculation were summarized in Table 3. Unit prices were based on market quotes of equipment manufacturers or production plants. TAC was calculated using the following formula:

$$\mathrm{TAC} = \mathrm{TOC} + \frac{\mathrm{TCC}}{\text{payback period}}$$

The capital costs of the distillation column were calculated based on the weight of the shell and the usage of packing material (details are provided in Supplementary Materials). The capital costs of heat exchangers were calculated based on the heat transfer area multiplied by the cost coefficient per unit area from market research.

Table 3. Basis and factors for economic analysis.

Columns (including manufacturing cost)	3733 USD/tonne
Column packing materials (BX gauze)	615 USD/m^3
Capital cost for heat exchanges	267 USD/m^2
Capital cost for steam compressors	2012 USD/kW
Capital cost for vacuum pumps	2051 USD/kW
Payback period	3 years
Electricity	0.1 USD/kWh
Cold water	0.13 USD/tonne
Steam	10.8 USD/GJ
Installation factor	1.3
Running time	8000 h/y
CO_2 emissions for steam	0.0896 kg/MJ
CO_2 emissions for electricity	0.674 kg/kWh (0.187 kg/MJ) [21]

The capital cost for steam compressors was more complicated. Although many researchers used empirical equations combined with price indices to estimate the cost of compressors [19,22,23], actual prices of steam compressors from commercial vendors in 2025 were significantly higher than values calculated from these empirical equations. To account for this discrepancy, compressors in this study used steam as the working medium and work under high temperatures (above 100 °C), leading to their prices being higher than common compressors. Therefore, after careful investigation of the price of steam compressors in the market, it was determined to use the correlation between power and cost of compressors to calculate equipment price [24–27]. The price factor obtained from market research is 2012 USD/kW (not including installation cost). Similar strategies were used to estimate prices of vacuum pumps, using a factor of 2051 USD/kW (not including installation cost).

The CO_2 emissions were calculated based on the amount of energy consumption, namely electricity and steam in this work. For electricity, according to a recent study [21], emission factors for electricity production in Xinjiang are 0.674 kg/kWh. For steam generation, the literature value of 0.0896 kg/MJ was used [21].

3. Process Design

3.1. Conventional Distillation Process

As shown in Table 2, crude phenols contained multiple xylenol and ethylphenol isomers, significantly increasing the complexity in distillation process. The conventional distillation process with four distillation columns is shown in Figure 1. The detailed parameters can be found in Table 4, and stream information (temperature, flow rates, composition, etc.) is provided in Supplementary information as Table S1. Design specifications to determine optimal reflux ratios and recovery rates for each column are provided in the Supplementary Materials. Due to the chemical instability and tendency to polymerization, the industrial practice for processing crude phenols is to avoid heating crude phenols over 170 °C. Therefore, the columns were operated under vacuum at pre-optimized pressure, so that the temperature of the column bottom liquid did not exceed 170 °C. The steam used for CD was saturated steam at 190 °C and 12.5 bar.

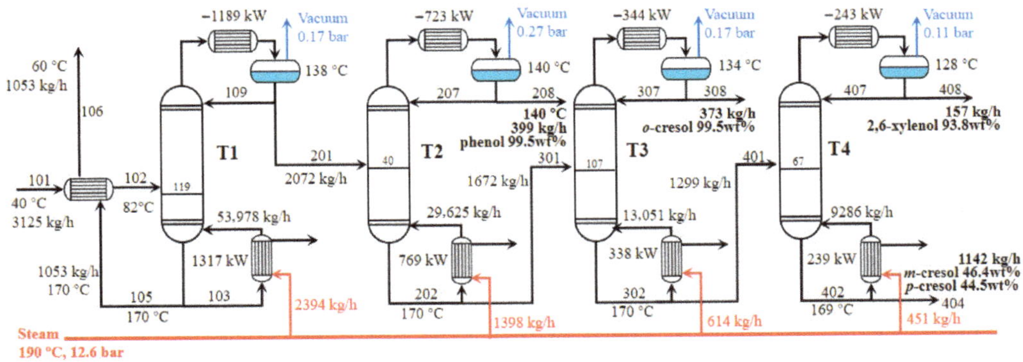

Figure 1. Design variables and parameters of the conventional distillation process for crude phenols.

Table 4. Design parameters of CD process [a].

Parameters	T1	T2	T3	T4
N	150	150	150	150
D/m	1.33	0.98	0.71	0.64
P_{top}/kPa	20	30	20	14
RR	3.13	11.22	5.98	12.16
M_{CW}/tonne·h^{-1}	128.7	78.2	37.2	26.0
A_{cond}/m^2	26.68	16.18	8.08	5.94
A_{reb}/m^2	155.88	95.22	44.08	28.36
P_{vac}/kW	3.37	1.81	1.77	2.01
TCC/10^6 USD	0.26	0.17	0.11	0.08

Table 4. Cont.

Parameters	T1	T2	T3	T4
$TOC/10^6$ USD·y^{-1}	0.65	0.38	0.17	0.13
$TAC/10^6$ USD·y^{-1}	0.74	0.44	0.21	0.15

[a] abbreviations: N, number of theoretical stages; D, diameter of column; P_{top}, pressure at the top stage of the column; RR, reflux ratio; M_{CW}, mass flow rate of cooling water; A_{cond}, heat transfer area of the condenser; A_{reb}, heat transfer area of the reboiler; P_{vac}, power of the vacuum pump.

Crude phenols were first fed into the first column T1 to remove heavy components, including all xylenols and ethylphenols (except for o-ethylphenol and 2,6-xylenol due to their relatively low boiling points), with the bottom liquid exchanging heat with the feed stream. The top distillate after condensation entered the second column T2 to separate phenol from the mixture, and phenol of 99.5 wt% was obtained from the top distillate. The bottom liquid leaving T2 was subsequently fed into column T3 to obtain o-cresol of 99.5 wt% purity in the distillate. The remaining 2,6-xylenol, m-cresol, *p*-cresol and many other components were separated in the last column T4. A mixture of *m*-cresol and *p*-cresol, as well as 8.1 wt% o-ethylphenol, were discharged from the bottom, while 2,6-xylenol were mainly discharged from the top of the column.

3.2. Heat Pump-Assisted Distillation Process

Due to the small difference in boiling points of components and thus high reflux ratios in each column, the conventional refining process demands significant amounts of energy and steam (Figure 1), leading to a low commercial value of the process. Heat pump-assisted distillation (HPD), which could greatly reduce energy consumption and operation costs, was adopted for crude phenol separation. Typical HPD processes include three schemes: vapor recompression (VRC), bottom flash (BF), and external vapor compression (VC) [13]. Crude phenols are heat-sensitive, highly toxic, and strongly corrosive, and ask high requirements of the heat pump and sealing materials. The polymerization tendency of phenolic substances further limits the use of VRC and BF. The VC scheme uses steam as the cycling medium, which prevents the risk of blockage or corrosion during phenol vapor compression. Moreover, when the released heat from condenser is insufficient for reboilers, the utility system can supply additional heat for the steam, making the heat pump operation convenient and flexible. The stability and durability of steam compressors is also relatively mature at present, making VC the optimal scheme for crude phenol distillation.

The process of crude phenol refining using heat pump-assisted distillation with external vapor compression is shown in Figures 2 and 3, with the same design specifications. Because the heat duty of the condenser and reboiler in each column was not identical, a sink for hot water (90 °C), which can be found in many chemical plants, was used to flexibly withdraw and discharge excess hot water. Similarly, a small amount of makeup steam was also used for columns T1 and T2, as steam from heat pump was not sufficient for reboilers. For each column, hot water at 90 °C from the water sink exchanged heat with the vapor from the top of the column in the condenser, yielding steam and condensed distillate. The steam was then pressurized by steam compressors to increase temperature and pressure, and then mixed with makeup steam (or withdrawn excess steam). The adequate amount of steam entered the reboiler to vaporize liquid in the column kettle and was discharged to the hot water sink. As can be seen in Figures 2 and 3, the electricity input of compressors in each column is significantly lower than the heat duty required for the reboiler, which accounts for the reduced energy consumption in heat pump-assisted distillation compared to conventional distillation. For T1 and T2, a small amount of saturated steam was still necessary, while excess steam was generated for T3 and T4.

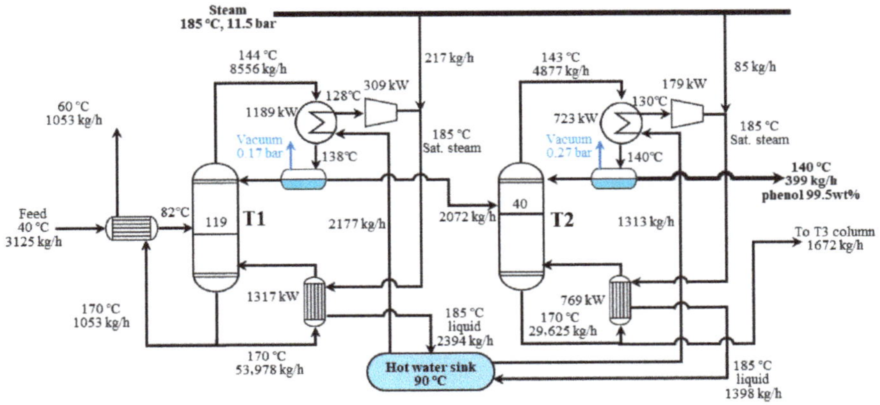

Figure 2. Design variables and parameters of HPD for crude phenol distillation (columns T1 and T2).

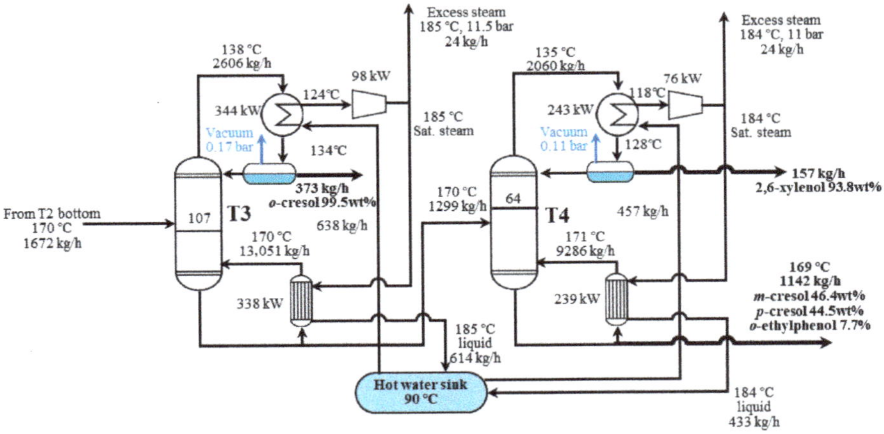

Figure 3. Design variables and parameters of heat pump-assisted distillation for crude phenol distillation (columns T3 and T4).

Compared to CD, HPD showed obvious advantages in terms of TAC, as shown in Table 5. For CD (Table 4), TAC was mainly contributed by operating cost, namely a large amount of pressurized steam for the reboilers, while capital costs for equipment such as columns or heat exchangers were low. For HPD, although capital costs (TCC) were increased due to expensive compressors, the much lower operating cost (TOC) still led to slightly lower TAC. It should also be noted that the cost calculations were based on the steam price of 10.8 USD/GJ and a payback period of three years. Other research has suggested higher steam prices and longer payback periods, and HPD would show larger advantages over CD in these cases. Another advantage of HPD over CD is the reduced consumption of cooling water. As seen in Table 4, CD used several hundred tonnes of cooling water per hour to cool down the top distillates. Although cooling water is readily obtained in most regions, some chemical plants located in arid areas may not have access to such a large amount of inexpensive cooling water. In this respect, HPD may be a desirable alternative over CD for separation processes.

Table 5. Design parameters of HPD process [a].

Parameters	T1	T2	T3	T4
N	150	150	150	150
D/m	1.33	0.98	0.71	0.64
P_{top}/kPa	20	30	20	14
RR	3.13	11.22	5.98	12.16
A_{cond}/m^2	204.0	141.3	63.0	42.2
A_{reb}/m^2	208.0	130.5	59.1	40.7
P_{vac}/kW	3.37	1.81	1.77	2.01
TCC/10^6 USD	1.15	0.70	0.39	0.30
TOC/10^6 USD·y^{-1}	0.30	0.16	0.08	0.06
TAC/10^6 USD·y^{-1}	0.68	0.40	0.21	0.16

[a] abbreviations: N, number of theoretical stages; D, diameter of column; P_{top}, pressure at the top stage of the column; RR, reflux ratio; A_{cond}, heat transfer area of the condenser; A_{reb}, heat transfer area of the reboiler; P_{vac}, power of the vacuum pump.

4. Sensitivity Analysis on Process Parameters

As shown above, crude phenol separation using HPD was designed, and preliminary economic performance was evaluated. To further optimize process variables, sensitivity analyses of multiple parameters in the HPD process were conducted in this section.

4.1. The Effect of Column Pressure

The operating pressure of the distillation columns affects the relative volatility of components, thereby affecting the separation difficulty, the required reflux ratio, energy consumption, and costs. The operating pressure also affects TCC, TAC, and TOC, while few studies deal with this topic. For crude phenols, according to industrial practice, vacuum in the columns was often preferred to keep the temperature below 170 °C to avoid decomposition and condensation. The sensitivity analysis of pressure on column costs was conducted to determine the optimum operating pressure. The pressure drop of each column was set constant to 8 kPa, while the pressure drop of condensers was set constant to 3 kPa. The feed stage was fixed and design specifications were satisfied, and the relationship between absolute pressure at the top of the column and reflux ratios is shown in Figure 4. With the increase in the top pressure of the T1 column, the relative volatility decreases, resulting in a gradual decrease in the column temperature difference (between top and bottom) and slightly higher separation difficulty, as the reflux ratio increased slightly from 3.05 to 3.17 with the pressure at the column top which increased from 8 kPa to 30 kPa. From the economic aspect, as shown in Figure 5, the decreased temperature difference between the top and bottom of the column would significantly decrease the required power for compression, leading to lower capital costs for compressors and lower TOC. The increase in column pressure would also lead to a decrease in capital costs of equipment and operation costs. The capital cost of column T1 shell reduced slightly with increasing pressure, as higher pressure implied higher vapor density and slower flow rate, leading to a slightly lower column diameter. The capital costs of heat exchangers and vacuum pumps changed slightly with pressure, as temperature differences were set to be constant. From an economic perspective, higher operating pressure in the column would indicate lower TCC, TAC, and TOC for the distillation process. Therefore, an intermediate pressure of 20 kPa was determined to be optimal so that temperature in the column kettle did not exceed 170 °C and decomposition or polymerization of crude phenols were not significant.

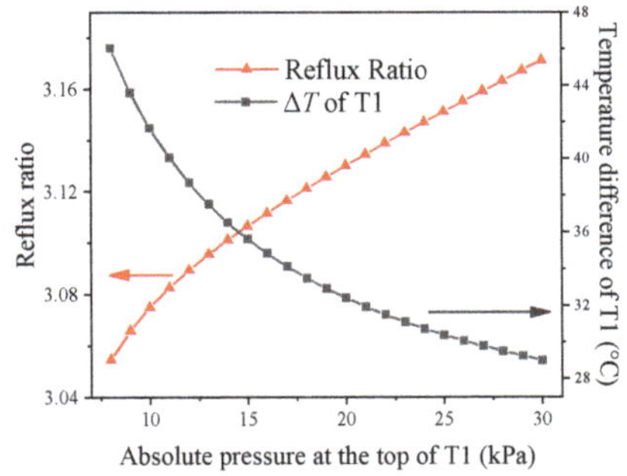

Figure 4. The reflux ratio and temperature difference between the top and bottom of T1 with respect to absolute pressure at the top of T1.

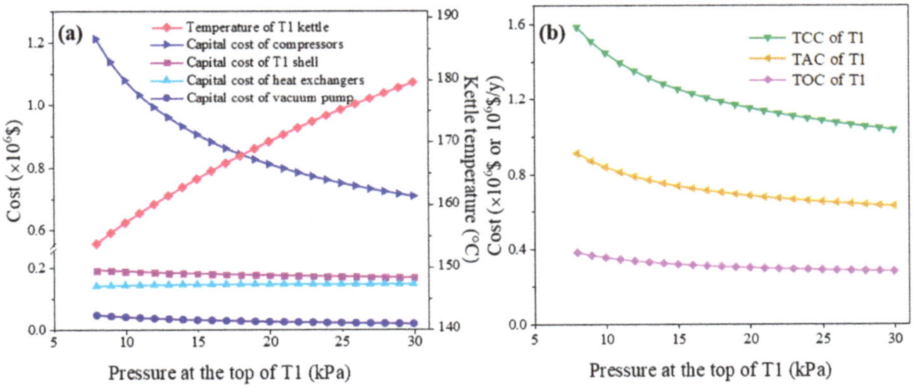

Figure 5. Sensitivity analysis of pressure at the top of T1 column on (**a**) costs of equipment and kettle temperature and (**b**) TCC, TAC, and TOC.

Also, it can be noticed in Figure 5a that the capital costs for compressors were much higher than for other equipment, such as column shells (including packing materials) or heat exchangers, which accounted for the majority of TCC and thus TAC. Although the application of HPD could reduce steam consumption, the capital costs for compressors still pose additional burdens on the profit outlook of the process. As shown later in this work, the profitability of HPD depends heavily on steam prices.

The sensitivity analysis of pressure on reflux ratios and TAC was also carried out on three other columns, and the results are shown in Figure 6. Regarding column T2 and T3, the increase in operating pressure led to slightly decreased reflux ratios and decreased TAC. Therefore, the optimum operating pressures for T2 and T3 were determined to be 30 kPa and 20 kPa, respectively, so that the temperature of the column kettle did not exceed 170 °C. For column T4, the reflux ratio increased more significantly with pressure, likely due to the similar boiling points of 2,6-xylenol and *m*-cresol/*p*-cresol, leading to smaller relative volatility between compounds and higher reflux ratios at increasing pressure. As a result of balancing the costs of heat exchangers and compressors, TAC of T4 column showed a concave curve at intermediate pressure. Accordingly, the pressure at the top of the T4 column was set to 14 kPa to achieve minimum TAC without exceeding 170 °C at the kettle.

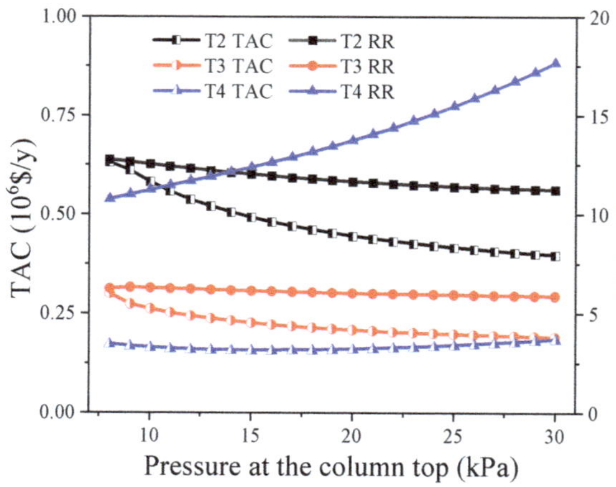

Figure 6. Sensitivity analysis of pressure at the top of column T2, T3, and T4 on TAC and reflux ratio (RR).

4.2. The Effect of Temperature Difference

The mechanism of a heat pump is to compress the vapor from the condenser and supply heat to the reboiler. In order to reduce the energy consumption of heat pumps, it is important to decrease the temperature gap between the condenser and reboiler to reduce the compression ratio and compression power. In general, for condensers, it is appropriate to set the temperature difference between cooling water and distillate to 10 °C. For reboilers, the temperature difference between heating steam and tower condensate is more crucial because an insufficient temperature difference may affect the stable operation of the distillation column. Therefore, it is necessary to analyze the effect of temperature difference between the heating steam and tower condensate on operation costs. Column T1 was used as an example, and the results are shown in Figure 7. The increased temperature difference would significantly lower the capital cost for heat exchangers because less heat transfer area would be required. The increase in temperature difference also leads to higher compression ratio and lower coefficient of performance (COP, defined by the ratio of useful process heat to the amount of primary energy supplied) [22], suggesting that a larger temperature difference would decrease compressor energy efficiency, which is in good agreement with the previous literature [27]. The capital cost for compressors increased as higher power was needed to lift the temperature of the steam, and operation cost for compressors would also increase. As a result, the total cost for equipment and TOC would also increase, suggesting that lower temperature difference would benefit the economic aspects of the HPD process. For reboilers, setting the temperature difference to 10 °C is relatively risky, as the simulation on the heavy components may be less accurate and incomplete vaporization may lead to poor separation. Therefore, it was determined to use 10 and 15 °C for the temperature difference in the condenser and reboiler, respectively.

4.3. The Effect of Theoretical Stages

The number of theoretical stages directly affects reflux ratios, thus affecting the equipment investment cost, operating cost, and TAC. Through sensitivity analysis, the changes in capital costs and operating costs were investigated when the number of theoretical plates varied from 100 to 150, while the feed stage was optimized to the optimal position simultaneously. Although the packed column height may also affect the pressure drop, due to the relatively low resistance of packing materials, the impact of the packing height on column pressure drop could be neglected. The results are shown in Figure 8. The increase in the theoretical stages of each column resulted in lower reflux ratios (Figure 8a) and higher capital costs of columns (Figure 8b), as expected. Lower reflux ratios further lead to lower operating costs and lower requirements for heat pump power (Figure 8c). Because the capital costs of compressors were much higher than column shells (including packing materials) and heat exchangers combined, the total annual costs (TAC) would tend to decrease with the increase in total number of stages (Figure 8d). Notably, although it is likely that increasing stage number would reduce the capital costs of equipment further, excessively high columns imply additional trouble in safety or construction. Therefore, it is appropriate to use 150 as the optimal value for the number of theoretical stages for four columns in this process to ensure the lowest TAC.

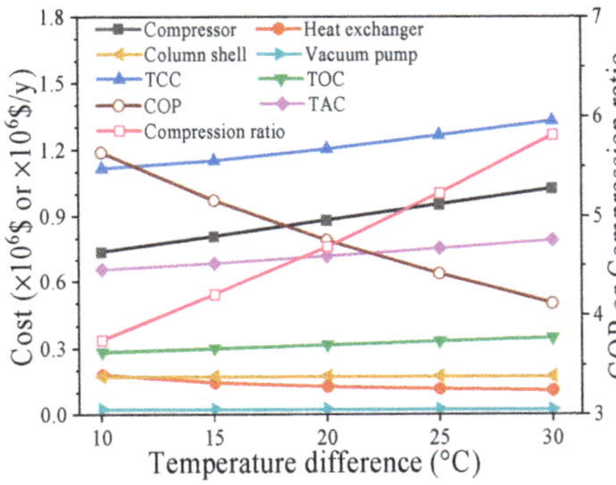

Figure 7. The effect of temperature difference in the reboiler on economic aspects of T1 column.

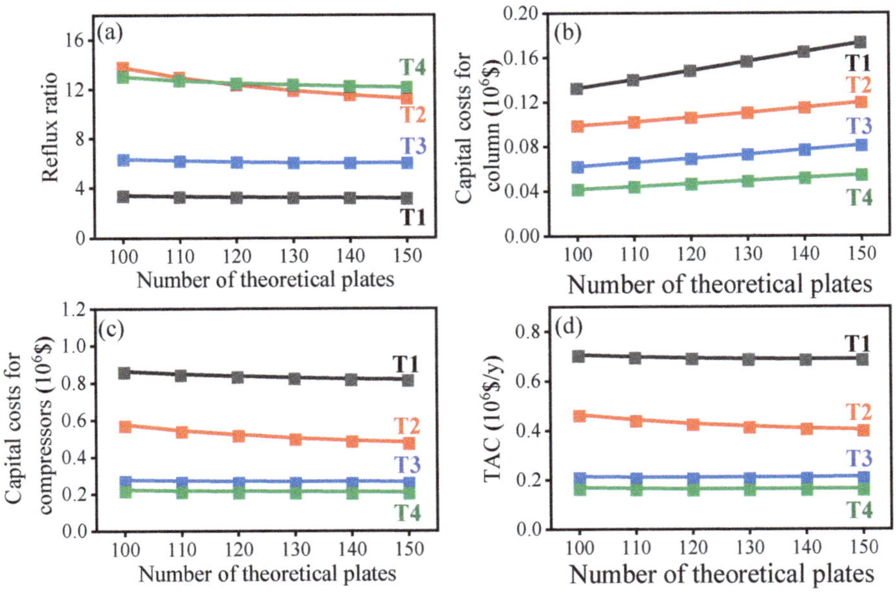

Figure 8. The effect of theoretical plate numbers on (**a**) reflux ratio, (**b**) capital costs of columns, (**c**) capital costs of steam compressors, and (**d**) TAC.

4.4. The Effect of Steam Cost on Process Economics

The major advantage of heat pump-assisted distillation (HPD) is using heat pumps to convert low-quality heat energy to high-quality heat energy to reduce steam consumption.

For scenarios where steam generation is inconvenient or relatively expensive, the immense heat required for reboilers can be supplied by vapor condensation at the expense of a relatively small amount of electricity. From the perspective of external energy consumption, HPD is undoubtedly energy efficient compared to conventional distillation (CD). However, HPD requires additional expenditure on compressors, which are more expensive than other equipment in the process, as shown in Figure 5a. If the steam price is relatively low, the reduced cost of steam consumption might not be able to compensate for the additional capital costs in heat pumps. In this respect, a threshold value for the steam price exists so that HPD is not competitive against CD in terms of TAC. To unveil such value, sensitivity analysis of steam price on process economics was conducted as shown in Figure 9, with steam price ranging from 3 to 21 USD/GJ. It can be noted that the operational cost responds strongly to steam price in CD, as CD consumes a large amount of steam in reboilers. For HPD, the energy input was mostly electricity and only a small amount of steam was used to compensate for the energy gap between supply and demand in reboilers, so the steam price barely affected the operation costs of HPD.

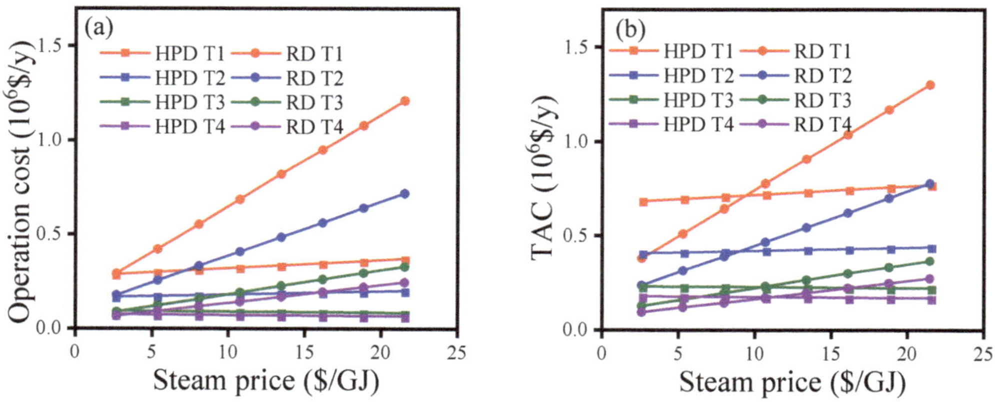

Figure 9. The effect of steam price on (**a**) operation cost and (**b**) TAC of each column in conventional distillation (CD) or heat pump-assisted distillation (HPD).

In terms of TAC, it is anticipated that the line for HPD and the line for CD intersect at an intermediate value. When the steam price was too high, the operation cost would exceed the capital cost for heat pump equipment, making HPD more competitive. Moreover, HPD also greatly reduced cooling water usage compared to CD, which also contributed to the reduced TAC. On the contrary, when steam was inexpensive, the capital costs for compressors could not be offset by reduced steam consumption in the payback period. According to this research, the threshold value lies within the range of 9 to 10.8 USD/GJ. When the steam price on site is higher than 10.8 USD/GJ, HPD is definitely more cost-effective than CD.

5. Environmental Impact

In addition to the reduced cost of energy, HPD shows great advantages over CD in terms of environmental impacts. The CO_2 emissions of HPD or CD can be calculated using factors, as shown in Table 3. According to the literature, the CO_2 emission factor for electricity varies depending on electricity generation or transmission [21]. In this study, the average emission factor for Xinjiang province (0.675 kg/kWh or 0.187 kg/MJ) was used. The CO_2 emission factor (0.0896 kg/MJ) for steam generation was adopted from the literature [28]. It can be observed from Figure 10 that CO_2 emission of HPD was less than half of that of CD. As shown in Tables 4 and 5, TOC, namely energy consumption (steam and electricity), in HPD was much lower than CD, leading to much lower CO_2 emission. Moreover, because the major energy consumption for HPD was electricity, it is anticipated that the reduction in CO_2 emission of HPD could be further reduced in the future with the development and expansion of renewable electricity.

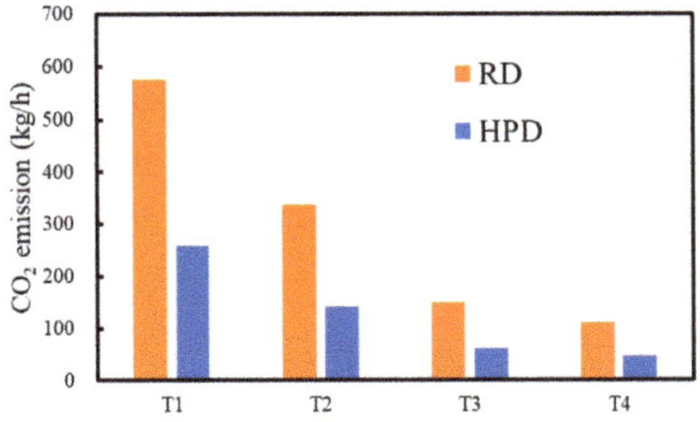

Figure 10. The amount of CO_2 emissions for HPD and CD.

6. Conclusions

Crude phenols are corrosive, heat-sensitive, and chemically unstable under high temperatures, which may cause blockages in equipment, pipelines, valves, and instruments. The application of heat pumps in crude phenol separation in previous research is not only scarce but also fails to consider practical issues that cannot be ignored during the crude phenol separation process. In this study, the practical and feasible separation of crude phenols was simulated using heat pump-assisted distillation (HPD), and a comprehensive techno-economic analysis was conducted. Due to their strong corrosiveness, high toxicity, and tendency to polymerize, crude phenols are only suitable for external vapor compression schemes. The effect of design parameters, such as column pressure, theoretical plate number, or temperature difference in heat exchangers, on process economics was investigated using process simulation software. Compared with conventional distillation (CD), the HPD process required extra expenditure on compressors but a much smaller amount of steam consumption, leading to its higher capital costs and lower operating cost. When the steam price exceeds 10.8 USD/GJ, HPD becomes advantageous over CD in terms of TAC. Moreover, HPD not only reduced steam consumption but also significantly reduced cooling water consumption, which is also important for regions where water was scarce. The carbon emissions generated by HPD are far lower than those of CD, significantly reducing its carbon footprint. With the rapid development and increasing share of renewable electricity, HPD has the potential to further reduce the CO_2 emissions and decrease operating costs. Although this is a simulative study and actual HPD might be less efficient than in the simulation, this work still provides insightful implications on key factors of the HPD process for crude phenol separation.

Supplementary Materials: The following supporting information can be downloaded at https://www.mdpi.com/article/10.3390/separations12110290/s1, Table S1: Design Specifications for distillation column simulation; Table S2: Stream information for conventional distillation process in Figure 1. Reference [29] are cited in the Supplementary Material.

Author Contributions: Conceptualization, Y.Z., J.X. and C.Z.; Methodology, D.M., L.Q. and J.Z.; Software, L.Q. and J.X.; Validation, J.Z. and C.Z.; Formal analysis, C.M.; Investigation, D.M., L.Q. and C.Y.; Resources, D.M. and J.X.; Data curation, C.M.; Writing—original draft, D.M.; Writing—review & editing, C.Z.; Visualization, C.Y.; Supervision, Y.Z. and J.X.; Project administration, Y.Z. and J.X.; Funding acquisition, Y.Z. All authors have read and agreed to the published version of the manuscript.

Funding: This research was funded by Xinjiang Uygur Autonomous Region Key R&D Program Project (2023B01015), Tiandi Co. Lit. Technology Innovation and Entrepreneurship Fund Special Project (2024-TD-QN001) and Technology Development Fund Project of CCTEG China Coal Research Institute (2024ZDI-03).

Data Availability Statement: The original contributions presented in this study are included in the article/Supplementary Materials. Further inquiries can be directed to the corresponding author(s).

Conflicts of Interest: Author D.M., L.Q., Y.Z., J.Z., and C.Y. were employed by the company CCTEG. The remaining authors declare that the research was conducted in the absence of any commercial or financial relationships that could be construed as a potential conflict of interest.

References

1. Richard, H.; Charles, G. Removal of Phenols from Phenol-Containing Stream. U.S. Patent 4256568A, 17 March 1981.
2. Weber, M.; Weber, M. Phenols. In *Phenolic Resins: A Century of Progress*; Pilato, L., Ed.; Springer: Berlin/Heidelberg, Germany, 2010; pp. 9–23.
3. Ma, Z.-H.; Li, S.; Dong, X.-Q.; Ju, C.-X.; Mu, D.; Yang, Z.; Li, M.; Guo, X.-H.; Zong, Z.-M.; Cong, X.-S.; et al. Value-added utilization of hydroxy-substituted aromatics in coal-derived liquids. *Fuel* **2023**, *340*, 127492. [CrossRef]
4. Bibi, A.; Bibi, S.; Abu-Dieyeh, M.; Al-Ghouti, M.A. Towards sustainable physiochemical and biological techniques for the remediation of phenol from wastewater: A review on current applications and removal mechanisms. *J. Clean. Prod.* **2023**, *417*, 137810. [CrossRef]
5. Zhang, C.; Liu, Y.; Gao, Z.; Huang, L.; Xiong, J. Process design, simulation and experimental studies of heteroazeotropic batch distillation for phenol dehydration. *Chem. Eng. Res. Des.* **2023**, *195*, 682–690. [CrossRef]
6. Yang, X.; Wang, B.; Luo, H.; Yan, S.; Dai, J.; Bai, Z. Efficient recovery of phenol from coal tar processing wastewater with tributylphosphane/diethyl carbonate/cyclohexane: Extraction cycle and mechanism study. *Chem. Eng. Res. Des.* **2020**, *157*, 104–113. [CrossRef]
7. Gao, J.; Li, Z.; Ding, C.; Sun, Z.; Ma, Y.; Xu, D.; Zhang, L.; Wang, Y. Separation of 1-methoxy-2-propanol and water by organic solvent extraction based on quantum chemistry calculation and liquid-liquid equilibrium experiment. *J. Chem. Thermodyn.* **2024**, *189*, 107201. [CrossRef]
8. Gao, P.; Zhang, Y.; Chen, H. Energy-efficient synthesis of crude phenol separation process using advanced heat integrated technology. *Energy Rep.* **2021**, *7*, 1888–1902. [CrossRef]
9. Meylemans, H.A.; Groshens, T.J.; Harvey, B.G. Synthesis of Renewable Bisphenols from Creosol. *Chemsuschem* **2012**, *5*, 206–210. [CrossRef]
10. Park, Y.; Hwang, S.; Won, S.; Kim, Y.; Hong, S.; Lee, J.; Lee, S.; Jeon, J.-K. Bisphenol F Synthesis from Formaldehyde and Phenol over Zeolite Y Extrudate Catalysts in a Catalyst Basket Reactor and a Fixed-Bed Reactor. *Catalysts* **2024**, *14*, 656. [CrossRef]
11. Xu, J.; Brodu, N.; Mignot, M.; Youssef, B.; Taouk, B. Synthesis and characterization of phenolic resins based on pyrolysis bio-oil separated by fractional condensation and water extraction. *Biomass Bioenergy* **2022**, *159*, 106393. [CrossRef]
12. Jana, A.K. Advances in heat pump assisted distillation column: A review. *Energ. Convers. Manag.* **2014**, *77*, 287–297. [CrossRef]
13. Kazemi, A.; Mehrabani-Zeinabad, A.; Beheshti, M. Recently developed heat pump assisted distillation configurations: A comparative study. *Appl. Energ.* **2018**, *211*, 1261–1281. [CrossRef]
14. Kiss, A.A.; Smith, R. Rethinking energy use in distillation processes for a more sustainable chemical industry. *Energy* **2020**, *203*, 117788. [CrossRef]
15. Pan, C.; Guo, J.; Feng, B.; Qiu, X.; Zhu, Q.; Song, H.; Gai, H.; Xiao, M.; Huang, T. Energy-saving extractive distillation process for the separation of close-boiling 2, 6-xylenol and p-cresol mixture. *J. Taiwan Inst. Chem. Eng.* **2024**, *159*, 105505. [CrossRef]
16. Luyben, W.L. Capital cost of compressors for conceptual design. *Chem. Eng. Process.-Process Intensif.* **2018**, *126*, 206–209. [CrossRef]
17. Liu, Q.; Zhang, X. Screening of Double Solvents Based on Multi-Index Evaluation Method for the Selective Separation m-Cresol from Model Oil. *ACS Omega* **2022**, *7*, 25798–25810. [CrossRef]
18. Haynes, W.M. *CRC Handbook of Chemistry and Physics*, 95th ed.; CRC Press: Boca Raton, FL, USA, 2014.

. Sun, S.; Yang, A.; Kong, Z.Y.; Huang, H.; Lv, L.; Zhou, Q.; Gu, B. The conceptual design and process intensification of the separation of ternary azeotropic mixture with heterogeneous characteristic. *Chem. Eng. Res. Des.* **2024**, *204*, 303–315. [CrossRef]
. Chen, Y.-C.; Hung, S.-K.; Lee, H.-Y.; Chien, I.L. Energy-Saving Designs for Separation of a Close-Boiling 1,2-Propanediol and Ethylene Glycol Mixture. *Ind. Eng. Chem. Res.* **2015**, *54*, 3828–3843. [CrossRef]
. Xiong, R.; Fu, C.; Chang, H.; Li, N.; Qu, S.; Zhao, D.; Xu, C.; Qi, J.; Xu, M. Emission factors and driving forces of provincial-level CO_2 from electricity production and consumption in China from 2013 to 2020. *J. Environ. Manag.* **2025**, *377*, 124644. [CrossRef]
2. Zhai, J.; Chen, X.; Sun, X.; Xie, H. Economically and thermodynamically efficient pressure-swing distillation with heat integration and heat pump techniques. *Appl. Therm. Eng.* **2023**, *218*, 119389. [CrossRef]
3. Zhang, Q.; Yang, S.; Shi, P.; Hou, W.; Zeng, A.; Ma, Y.; Yuan, X. Economically and thermodynamically efficient heat pump-assisted side-stream pressure-swing distillation arrangement for separating a maximum-boiling azeotrope. *Appl. Therm. Eng.* **2020**, *173*, 115228. [CrossRef]
4. Schlosser, F.; Jesper, M.; Vogelsang, J.; Walmsley, T.G.; Arpagaus, C.; Hesselbach, J. Large-scale heat pumps: Applications, performance, economic feasibility and industrial integration. *Renew. Sustain. Energy Rev.* **2020**, *133*, 110219. [CrossRef]
5. Meyers, S.; Schmitt, B.; Vajen, K. The future of low carbon industrial process heat: A comparison between solar thermal and heat pumps. *Sol. Energy* **2018**, *173*, 893–904. [CrossRef]
6. Arpagaus, C.; Bless, F.; Uhlmann, M.; Schiffmann, J.; Bertsch, S.S. High temperature heat pumps: Market overview, state of the art, research status, refrigerants, and application potentials. *Energy* **2018**, *152*, 985–1010. [CrossRef]
7. Vieren, E.; Demeester, T.; Beyne, W.; Magni, C.; Abedini, H.; Arpagaus, C.; Bertsch, S.; Arteconi, A.; De Paepe, M.; Lecompte, S. The Potential of Vapor Compression Heat Pumps Supplying Process Heat between 100 and 200 °C in the Chemical Industry. *Energies* **2023**, *16*, 6473. [CrossRef]
8. You, X.; Rodriguez-Donis, I.; Gerbaud, V. Reducing process cost and CO2 emissions for extractive distillation by double-effect heat integration and mechanical heat pump. *Appl. Energy* **2016**, *166*, 128–140. [CrossRef]
9. *GB/T 25198-2023*; Pressure Vessel Heads. Standardization Administration of the People's Republic of China: Beijing, China, 2023.

separations

Article

Vapor Liquid Equilibrium Measurement and Distillation Simulation for Azeotropic Distillation Separation of H_2O/EM Azeotrope

Chunli Li [1], Jinxin Zhang [1], Jiqing Rao [1], Kaile Shi [1], Yuze Sun [1], Wen Liu [2,*] and Jiapeng Liu [1,*]

1 National-Local Joint Engineering Laboratory for Energy Conservation in Chemical Process Integration and Resources Utilization, School of Chemical Engineering and Technology, Hebei University of Technology, Tianjin 300130, China
2 College of Basic Science, Tianjin Agricultural University, Tianjin 300392, China
* Correspondence: liuwen@tjau.edu.cn (W.L.); liujiapeng@hebut.edu.cn (J.L.)

Abstract: Since H_2O and Ethylene Glycol Monomethyl Ether (EM) form a minimum-boiling azeotrope, 1-pentanol, 1-hexanol, and 1-heptanol are selected as entrainers to separate the azeotropic mixture (H_2O/EM) using azeotropic distillation. The binary vapor liquid equilibrium (VLE) data were determined at 101.3 kPa, including H_2O/EM, EM/1-pentanol, EM/1-hexanol, EM/1-heptanol, H_2O/1-pentanol, H_2O/1-hexanol and H_2O/1-heptanol. Meanwhile, the Herington area test was used to validate the thermodynamic consistency of the experimental binary data. The VLE data for the experimental binary system were analyzed using the NRTL, UNIQUAC, and Wilson activity coefficient models, showing excellent agreement between predictions and measurements. Finally, molecular simulations were employed to calculate interaction energies between components, providing insights into the VLE behavior. The azeotropic distillation process was simulated using Aspen Plus to evaluate the separation performance and determine the optimal operating parameters. Therefore, this study provides guidance and a foundational basis for the separation of H_2O/EM systems at 101.3 kPa.

Keywords: vapor liquid equilibrium; azeotrope; ethylene glycol monomethyl ether; Herington area test; NRTL

1. Introduction

EM, also known as methyl cellosolve, is an important chemical raw material [1], Due to the presence of both hydrophobic and hydrophilic functional groups in this compound, it is widely utilized in diverse industrial and commercial applications, including household products, paints, inks [2], coatings, cleaning solutions and biochemical applications [1,3]. Since the H_2O/EM system forms a completely miscible binary azeotrope, specialized distillation methods must be employed for separation, including extractive distillation [4,5], azeotropic distillation [6,7], and salting-out distillation [8]. In this study, 1-pentanol, 1-hexanol, and 1-heptanol were selected as azeotropic agents to separate the H_2O/EM azeotropic mixture. The advantages of azeotropic distillation in this separation process lie in its ability to break azeotropic limitations, technological maturity, and high purity of the final product. Compared to alternative technologies such as extraction and membrane separation, azeotropic distillation demonstrates strong competitiveness due to its reliability and high product purity.

Currently, there are no experimental reports on VLE data for entrainer screening in the H_2O/EM azeotropic system. To separate the azeotropic mixture using azeotropic distillation and selected entrainers, binary isobaric VLE data for the systems are required (H_2O/EM, EM/1-pentanol, EM/1-hexanol, EM/1-heptanol, H_2O/1-pentanol, H_2O/1-hexanol and H_2O/1-heptanol). Numerous studies have reported binary VLE data for the H_2O/EM azeotropic system. Bejarano [9] studied the system under varying pressures (74.5 kPa, 101.3 kPa, and 134.0 kPa). Ochi K [10] reported data for the H_2O/1-hexanol system under the 101.3 kPa, Krishnaiah reported the vapor liquid equilibrium data for EM/1-pentanol at 40 kPa and 95 kPa [11], and in NIST, there are binary VLE data for H_2O/1-pentanol at 101.3 kPa. Notably, the NIST database currently lacks data for the following binary systems at 101.3 kPa (EM/1-pentanol, EM/1-hexanol, EM/1-heptanol, and H_2O/1-heptanol).

In this work, binary VLE data were measured at 101.3 kPa for the following systems (H_2O/EM, EM/1-pentanol, EM/1-hexanol, EM/1-heptanol, H_2O/1-pentanol, H_2O/1-hexanol and H_2O/1-heptanol). The measured binary VLE data were validated for thermodynamic consistency using the Herington area test [12]. The experimental data were correlated with the Non-Random Two-Liquid (NRTL) [13], Universal Quasi-Chemical (UNIQUAC) [14], and Wilson activity coefficient models [15,16]. The azeotropic distillation process was simulated using Aspen Plus to evaluate the separation performance and determine the optimal operating parameters.

2. Experiment

2.1. Chemicals

A summary of the data for these chemicals is given in Table 1. All chemicals used in this study were used without further purification.

Table 1. Details of the chemicals.

Chemical	Molecular Formula	CAS No.	Mass Fraction	Purification Method	Source	Analysis Method
ethylene glycol monomethyl ether	$C_3H_8O_2$	109-86-4	>0.995	none	Shandong Keyuan Biochemical Co., Ltd. (Heze, China)	GC
1-pentanol	$C_5H_{12}O$	71-41-0	>0.990	none	Shandong Keyuan Biochemical Co., Ltd.	GC
1-hexanol	$C_6H_{14}O$	111-27-3	>0.990	none	Shanghai Meryer Co., Ltd. (Shanghai, China)	GC
1-heptanol	$C_7H_{16}O$	111-70-6	>0.990	none	Shanghai Meryer Co., Ltd.	GC
water	H_2O	7732-18-5	>0.990	none	Hangzhou Wahaha Group Co., Ltd. (Hangzhou, China)	GC

2.2. Apparatus and Experimental Procedure for VLE Data Collection

The binary VLE data were determined using a modified Othmer distill [17,18], As shown in Figure 1. The effectiveness of the entire experimental setup was validated through previously conducted and published studies [19]. Prior to the experiments, the temperature and pressure measurement system of the Othmer boiling point apparatus was calibrated using a standard substance (H_2O). Subsequently, standard samples of known concentration were prepared, and a GC analytical calibration curve was established. VLE experiments were then conducted for verification to ensure the reliability of the experimental setup. As seen in Figure 2, the experimental data show good agreement with literature values [9] under comparable conditions, providing an independent validation of existing data. The experimental data will be further fitted in the following sections.

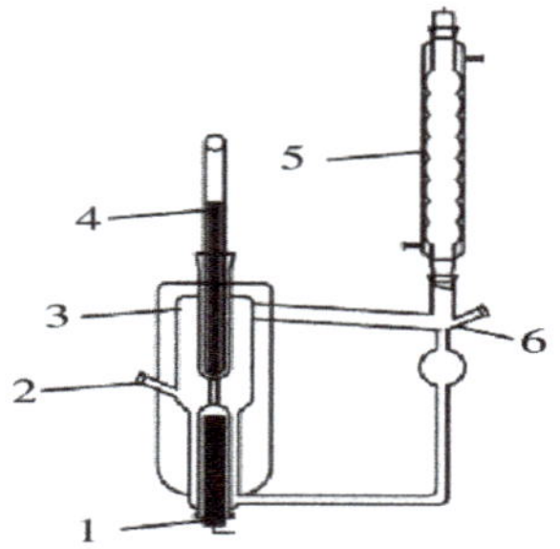

Figure 1. Schematic diagram of the modified Othmer distill experimental setup. 1: heating rod; 2: liquid-phase sampling point; 3: equilibrium chamber; 4: thermometer; 5: condenser; 6: vapor-phase sampling point.

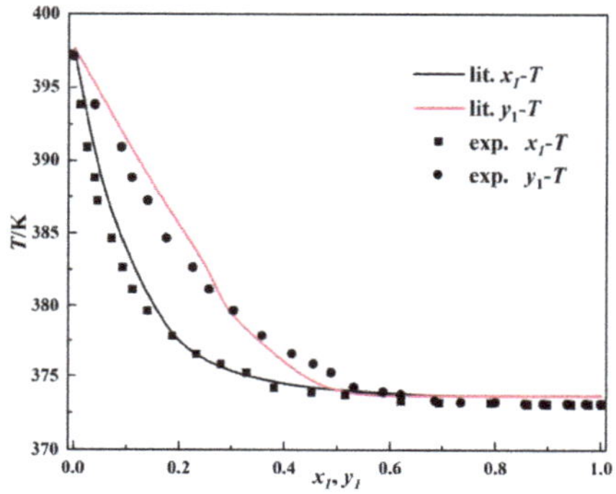

Figure 2. Vapor liquid equilibrium phase diagram and experimental data of H_2O/EM system.

During measurements, temperature fluctuations of the mixture remained within 0.10 K over approximately 100 min, confirming the attainment of equilibrium. In the procedure, the mixture in the equilibrium still was heated at the bottom to generate vapor, which was then condensed in a spherical condenser. The condensed liquid flowed back into the still, where it was reheated to boiling, cycling repeatedly until equilibrium was achieved. After the system stabilized at equilibrium, liquid-phase and vapor-phase samples were extracted using microsyringes for chromatographic analysis.

2.3. *Experimental Procedures*

The vapor-phase and liquid-phase samples were analyzed by gas chromatography (GC), with an injection volume of 0.4 μL for each analysis. The gas chromatographic analytical conditions are listed in Table 2.

Table 2. Details of GC.

Project	Characteristic	Description
Column	Type	Packing column
	Specification	Porapak Q (3 mm × 2 m)
Carrier gas	Specification	H_2 (99.999%, mass fraction)
	Flow rate	40 cm^3/min
Detector	Type	Thermal conductivity detector (TCD)
	Temperature	523 K

Table 2. *Cont.*

Project	Characteristic	Description
Injector	Temperature	523 K
Column	Temperature	503 K

2.4. Molecular Simulation Methods

As a pivotal computational method, molecular simulation uncovers the fundamental properties and dynamic processes of substances at the atomic level, providing crucial insights into mechanisms that underpin material behavior and complementing experimental observations. Its essence lies in reconstructing the dynamic processes of the microscopic world through mathematical models and algorithms. Molecular modeling facilitates the simulation of experimentally inaccessible information, including inter-molecular interactions and molecular dynamics trajectories. This method has been applied to studies of adsorption processes [20,21], crystallization processes [22,23], and phase equilibrium processes [24,25]. Interaction energies among system components can be quantified using molecular simulations that employ density functional theory.

The theoretical quantum chemical calculations were performed using the Gaussian 16 software package. Geometry optimization and subsequent vibrational frequency calculations for the studied complexes were performed at the B3LYP-D3BJ/TZVP level. The optimized geometries were established as local minima on the potential energy surface (PES) through vibrational frequency analysis, which confirmed the absence of imaginary frequencies, thereby excluding saddle points or transition states. The B3LYP-D3BJ/TZVP [26] level of theory has been widely validated as a reliable approach for predicting molecular electrostatic potentials and solvation surface charge distributions. All quantum chemical calculations in this study were performed under gas-phase conditions. While this method efficiently provides molecular electronic structures and optimized geometries, it is crucial to emphasize its inherent limitation: it does not account for solvation effects in liquid-phase environments. The interaction energy (denoted as E_{inter}), a fundamental measure of binding affinity in the optimized complexes, was evaluated according to the standard definition [25]. It is calculated as the energy difference between the complex and its constituent monomers at their optimized geometries within the complex (Equation (1)) [26]:

$$E_{inter} = E_{AB} - E_A - E_B . \tag{1}$$

Among them, E_{AB}, E_A, and E_B represent the total energies of the complex (A+B), monomer A, and monomer B, respectively. The energy mixing (ΔE) is related to the overheating enthalpy, which can be defined as the difference between the interaction energy of mixed molecules and the interaction energy of pure components. To model the energetic consequences of intermolecular interactions upon mixing, Equation (2) provides the fundamental framework for computing the system's energy perturbation:

$$\Delta E = 2E_{AB} - E_{AA} - E_{BB}. \tag{2}$$

If the calculated value of ΔE is zero, the system is considered athermal (ideal mixing). If there is a significant deviation from zero, it indicates a greater deviation from ideality.

3. Experimental Results and Discussion

3.1. VLE Data of Binary Systems

Under low-pressure or medium-pressure conditions, the expressions for VLE [27,28] are as follows:

$$\gamma_i = \frac{P y_i}{P_i^s x_i}. \tag{3}$$

Among them, x_i and y_i represent the mole fractions of component i in the liquid and vapor phases, respectively, P denotes the system pressure, and P_i^s is the saturation vapor pressure of pure component i, which can be calculated using the Extended Antoine Equation [29] from Aspen V12 Plus. Compared to the standard Antoine equation [30], the additional parameters introduced in the Extended Antoine Equation enhance its flexibility and allow it to describe the entire vapor pressure curve. Table 3 lists the Antoine parameters for C_1 to C_9. Extended Antoine Equation Calculation:

$$lnp_i = C_{1i} + \frac{C_{2i}}{T + C_{3i}} + C_{4i} + C_{5i} lnT + C_{6i} T^{C^{7i}} for\ C_{8i} \le T \le C_{9i}. \tag{4}$$

Among them, T (K) represents the system temperature, and P (kPa) denotes the system pressure. Based on the calculations using the above formulas and parameters, Tables 4–10 present the experimentally obtained binary VLE data and activity coefficients (γi). As observed in Tables 8–10, activity coefficients (γ) exhibit abnormally high values in some cases. This can be attributed to the significant difference between the strong hydrogen bonding among H_2O molecules and the pronounced hydrophobicity of the long alkyl chain of alcohol molecules. This leads directly to highly non-ideal behavior in the liquid phase and a strong tendency for liquid-phase splitting (partial miscibility). In dilute aqueous solutions, alcohol molecules exhibit extremely high fugacity, indicating a substantial tendency to "escape" from the liquid phase, which is reflected as activity coefficients γ >> 1.

Table 3. Parameters for the extended Antoine equation.

Component	C_{1i}	C_{2i}	C_3	C_{4i}	C_{5i}	C_{6i}	C_{7i}	C_{8i}	C_{9i}
EM	62.6982	−7444.7	0	0	−6.581	8.883×10^{-18}	6	188.15	597.6
1-pentanol	107.842	−10,643	0	0	−12.858	1.249×10^{-17}	6	195.56	588.1
1-hexanol	128.512	−12,288	0	0	−15.732	1.270×10^{-17}	6	228.55	611.3
1-heptanol	140.5020	−13,466	0	0	−17.353	1.128×10^{-17}	6	239.15	632.3
H_2O	66.7412	−7258.2	0	0	−7.304	4.165×10^{-6}	2	273.16	647.1

Table 4. Binary VLE data for H_2O (1)/EM (2).

T/K	x_1	y_1	γ_1	γ_2
373.15	1.0000	1.0000	1.0025	-
373.15	0.9759	0.9719	0.9985	2.6402
373.15	0.9426	0.9367	0.9962	2.5050
373.15	0.8969	0.8914	0.9963	2.3911
373.15	0.8602	0.8564	0.9980	2.3319
373.25	0.7906	0.7999	1.0107	2.1606
373.25	0.6930	0.7348	1.0591	1.9538
373.35	0.6202	0.6862	1.1014	1.8616
373.75	0.5148	0.6206	1.1831	1.7366
373.95	0.4514	0.5870	1.2672	1.6599
374.25	0.3800	0.5315	1.3486	1.6482
375.25	0.3286	0.4884	1.3832	1.6036

Table 4. *Cont.*

T/K	x_1	y_1	γ_1	γ_2
375.85	0.2791	0.4548	1.4845	1.5581
376.55	0.2323	0.4142	1.5856	1.5333
377.85	0.1866	0.3576	1.6286	1.5158
379.65	0.1393	0.3037	1.7411	1.4582
381.15	0.1110	0.2563	1.7523	1.4316
382.65	0.0925	0.2257	1.7604	1.3870
384.65	0.0716	0.1753	1.6520	1.3494
387.25	0.0444	0.1401	1.9513	1.2535
388.85	0.0391	0.1097	1.6492	1.2242
390.95	0.0247	0.0897	1.9957	1.1517
393.85	0.0117	0.0395	1.6871	1.0928
397.15	0.0000	0.0000	-	1.0134

Table 5. Binary VLE data for EM (1)/1-pentanol (2).

T/K	x_1	y_1	γ_1	γ_2
410.65	0.0000	0.0000	-	1.0101
408.45	0.0683	0.1099	1.1609	1.0374
406.95	0.1363	0.1815	1.0038	1.0818
405.35	0.2053	0.3082	1.1857	1.0488
404.15	0.2942	0.4081	1.1357	1.0524
402.05	0.4364	0.5607	1.1202	1.0516
400.85	0.5371	0.6438	1.0839	1.0823
400.25	0.5924	0.6909	1.0742	1.0891
399.85	0.6641	0.7482	1.0505	1.0918
398.95	0.7639	0.8215	1.0307	1.1370
398.75	0.7962	0.8452	1.0237	1.1505
398.15	0.8249	0.8749	1.0420	1.1055
397.15	1.0000	1.0000	1.0134	-

Table 6. Binary VLE data for EM (1)/1-hexanol (2).

T/K	x_1	y_1	γ_1	γ_2
430.15	0.0000	0.0000	-	0.9936
424.65	0.0874	0.2152	1.1321	1.0164
419.75	0.1732	0.4001	1.2116	1.0065
414.55	0.2594	0.5148	1.2020	1.0839
411.05	0.3742	0.6295	1.1253	1.1069
407.95	0.4508	0.7147	1.1599	1.0850
406.35	0.5161	0.7522	1.1177	1.1335
405.05	0.5730	0.7894	1.0979	1.1450
403.85	0.6256	0.8268	1.0915	1.1232
402.95	0.6805	0.8583	1.0701	1.1138
401.65	0.7422	0.8886	1.0566	1.1390
400.65	0.8029	0.9124	1.0338	1.2169
399.75	0.8483	0.9363	1.0323	1.1888
397.15	1.0000	1.0000	1.0134	-

Table 7. Binary VLE data for EM (1)/1-heptanol (2).

T/K	x_1	y_1	γ_1	γ_2
448.15	0.0000	0.0000	-	1.0153
438.15	0.0497	0.2507	1.6423	1.0869
431.95	0.1003	0.4040	1.5310	1.1158
423.75	0.1921	0.5930	1.4533	1.1212
417.15	0.2890	0.7399	1.4423	1.0310
412.15	0.3839	0.8284	1.3986	0.9461
408.15	0.4695	0.8895	1.3780	0.8254
406.25	0.5215	0.9289	1.3701	0.6344
403.65	0.6005	0.9408	1.3019	0.7020
402.85	0.6307	0.9462	1.2768	0.7131
401.65	0.6801	0.9517	1.2350	0.6153
397.15	1.0000	1.0000	1.0134	-

Table 8. Binary VLE data for H_2O (1)/1-pentanol (2).

T/K	x_1	y_1	γ_1	γ_2
411.15	0.0000	0.0000	-	0.9937
381.55	0.0272	0.3064	8.4329	2.0641
379.55	0.0430	0.4375	8.1540	1.8444
374.15	0.0764	0.4701	5.9515	2.2514
371.95	0.1032	0.5569	5.6443	2.1293
369.25	0.1381	0.5553	4.6402	2.4990
369.15	0.1499	0.5798	4.4790	2.4050
369.15	0.1508	0.5690	4.3686	2.4696
369.15	0.2226	0.5696	2.9639	2.6933
369.15	0.3074	0.5641	2.1253	3.0622
368.75	0.4183	0.5823	1.6358	3.5559
368.75	0.5180	0.5772	1.6358	3.5559
368.75	0.5743	0.5517	1.1291	5.2140
368.75	0.6948	0.5673	0.9597	7.0187
369.05	0.7991	0.5880	0.8554	10.0224
369.05	0.9480	0.5600	0.6867	41.3464
373.15	1.0000	1.0000	1.0025	-

Table 9. Binary VLE data for H_2O (1)/1-hexanol (2).

T/K	x_1	y_1	γ_1	γ_2
429.65	0.0000	0.0000	-	1.0092
382.15	0.0507	0.6301	9.1174	2.2674
373.35	0.0988	0.6500	6.5472	3.3602
371.85	0.2226	0.6862	3.2371	3.7465
370.85	0.2985	0.6805	2.4819	4.4326
370.75	0.4211	0.6933	1.7991	5.1807
370.75	0.4981	0.6850	1.5027	6.1370
370.75	0.6091	0.6750	1.2110	8.1298
370.85	0.7271	0.7059	1.0570	10.4880
370.75	0.7558	0.6904	0.9981	12.3998
370.75	0.8713	0.7032	0.8820	22.5461
370.95	0.9790	0.6840	0.7580	145.7353
373.15	1.0000	1.0000	1.0025	-

Table 10. Binary VLE data for H_2O (1)/1-heptanol (2).

T/K	x_1	y_1	γ_1	γ_2
448.15	0.0000	0.0000	-	1.0153
377.45	0.0479	0.8020	14.4234	3.1665
374.65	0.1002	0.8034	7.6239	3.8227
372.25	0.1499	0.8032	5.5486	4.5727
372.05	0.2001	0.8285	4.3168	4.2776
372.05	0.2926	0.8352	2.9760	4.6481
371.85	0.4253	0.8392	2.0725	5.6389
372.15	0.5162	0.8429	1.6965	6.4471
371.95	0.6230	0.8310	1.3960	8.9871
372.15	0.6868	0.8301	1.2560	10.7639
371.95	0.7911	0.8359	1.1058	15.7493
372.15	0.9780	0.8400	0.8924	144.3722
373.15	1.0000	1.0000	1.0025	-

3.2. Thermodynamic Consistency Test for the Experimental Data (Herington Integral Method)

To ensure the accuracy and reliability of the measured VLE data, thermodynamic consistency tests must be performed on the experimental data. Common methods include the integral test, differential test, and slope method. To assess adherence to the Gibbs-Duhem equation, the thermodynamic consistency of the experimental phase equilibrium data was rigorously evaluated via Herington area test [31]. The Herington method is a semi-empirical approach that involves plotting x_1 against $\ln(\gamma_1/\gamma_2)$ (as shown in Figure 3) and defining and examining the relationship between D and J. The specific steps are as follows:

$$D = \frac{\int_{x_i=0}^{x_i=1} ln\frac{\gamma_2}{\gamma_1}dx_i}{\int_{x_i=0}^{x_i=1} \left|ln\frac{\gamma_2}{\gamma_1}\right|dx_i} \times 100, \quad (5)$$

$$J = 150 \times \frac{T_{max} - T_{min}}{T_{min}}. \quad (6)$$

Among them, T_{max} (K) corresponds to the maximum temperature of the system. T_{min} (K) corresponds to the minimum temperature of the system.

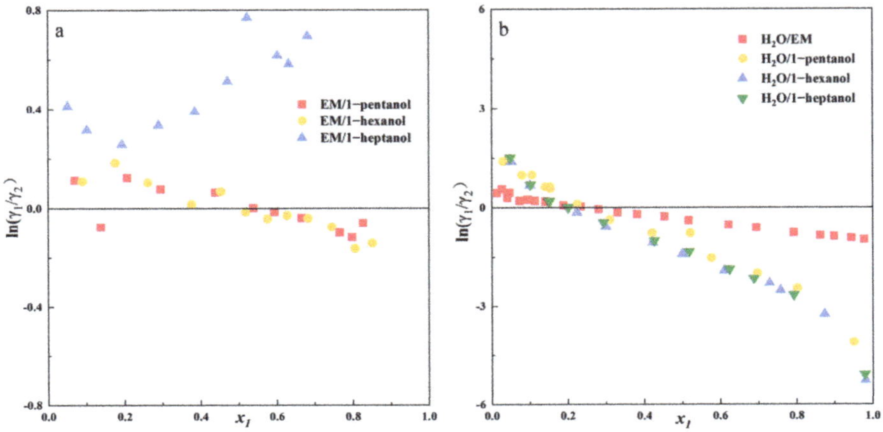

Figure 3. ln (γ_1/γ_2) ~ x_1 diagram. (**a**): ■ EM/1-pentanol; • EM/1-hexanol; ▲ EM/1-heptanol; (**b**): ■ H_2O/EM; • H_2O/1-pentanol; ▲ H_2O/1-hexanol; ▼ H_2O/1-heptanol.

The calculation results are presented in Table 11, $D - J$ represents the criterion for evaluating thermodynamic consistency using the Herington integral test method. If $|D - J| < 10$, the data exhibits reasonable reliability. As shown in Table 11, the measured VLE data for the binary system demonstrate reliable consistency.

Table 11. Result of Herrington integral test.

Binary System	$\vert D - J\vert$
H_2O/EM	5.9450
EM/1-pentanol	6.6655
EM/1-hexanol	2.8024
EM/1-heptanol	2.4528
H_2O/1-pentanol	8.0924
H_2O/1-hexanol	8.8198
H_2O/1-heptanol	8.1581

3.3. Correlation of Binary VLE Data

To validate experimental data, the VLE data obtained from experiments were correlated using the NRTL, UNIQUAC, and Wilson activity coefficient models, which appear in Equations (7)–(9). The calculated correlated binary interaction parameters are listed in Table 12, and a comparison between experimental and predicted data is shown in Figure 4. To quantify model prediction fidelity, root mean square deviations (RMSD) of vapor-phase mole fractions and equilibrium temperatures were computed via Equations (10) and (11), serving as key metrics for thermodynamic model validity.

NRTL:

$$\ln\gamma_i = \frac{\sum_{j=1}^{n} x_{ji}G_{ji}x_j}{\sum_{k=1}^{n} G_{ki}x_k} + \sum_{j=1}^{n}\left(\frac{x_jG_{ij}}{\sum_{k=1}^{n} G_{kj}x_k}\right)\left(\tau_{ij} - \frac{\sum_{l=1}^{n} \tau_{lj}G_{lj}x_l}{\sum_{k=1}^{n} G_{kj}x_k}\right), \tag{7}$$

where $\tau_{ij} = a_{ij} + \frac{b_{ij}}{T}; G_{ij} = \exp(-\alpha_{ij}\tau_{ij})$.

UNIQUAC:

$$\ln\gamma_i = \ln\frac{\phi_i}{x_i} + \frac{z}{2}q_i\ln\frac{\theta_i}{\phi_i} + l_i + \frac{\phi_i}{x_i}\sum_{j=1}^{n} x_jl_j + q_i^t\left(1 - \ln t_i^t - \frac{\sum_{j=1}^{n} \theta_j\tau_{ij}}{t_j^t}\right), \tag{8}$$

where $\tau_{ij} = \exp\left(a_{ij}+\frac{b_{ij}}{T}\right)$.

Wilson:

$$\ln\gamma_i = 1 - \ln\left(\sum_{j=1}^{n} A_{ij}x_j\right) - \sum_{k=1}^{n}\left(\frac{x_kA_{ki}}{\sum_{j=1}^{n} A_{kj}x_j}\right), \tag{9}$$

where $A_{ij} = a_{ij} + \frac{b_{ij}}{T}$.

$$\mathrm{RMSD}(\mathrm{y_i}) = \sqrt{\frac{\sum_i^N \left(y_{i,exp} - y_{i,cal}\right)^2}{N}} \tag{10}$$

$$\mathrm{RMSD}(T_\mathrm{i}) = \sqrt{\frac{\sum_i^N \left(T_{i,exp} - T_{i,cal}\right)^2}{N}} \tag{11}$$

Table 12. The correlated binary interaction parameters (a_{ij}; a_{ji}; b_{ij}; b_{ji}), standard deviations of vapor compositions (δ_{y1}) and temperatures (δ_T).

System	Model	Correlation Paramenters				RMSD	
		a_{ij}	a_{ji}	b_{ij}/K	b_{ji}/K	δ_{y1}	δ_T
	NRTL	−13.7500	6.3087	5804.2600	−2612.2645	0.0076	0.3232
H_2O/EM	UNIQUAC	1.1661	0.1903	−141.1685	−619.4664	0.0067	0.3062
	WILSON	−3.3899	9.4101	1387.1798	−4020.3589	0.0067	0.3117
	NRTL	2.9908	−0.1247	−916.6271	−100.2345	0.0097	0.2624
EM/1-pentanol	UNIQUAC	−1.5565	0.5743	470.7308	−125.3232	0.0097	0.2610
	WILSON	0.01484	−2.6222	159.0824	742.3043	0.0097	0.2623
	NRTL	11.9088	−8.4514	−4512.7373	3221.0569	0.0112	0.3033
EM/1-hexanol	UNIQUAC	−4.4627	4.7822	1812.4600	−1970.3248	0.0116	0.3105
	WILSON	7.07436	−9.9947	−2752.0434	3841.7263	0.0114	0.3072
	NRTL	29.5548	13.9507	−10000.0000	−6383.2185	0.0104	0.1710
EM/1-heptanol	UNIQUAC	−25.5473	−1.65054	9051.5620	1090.3780	0.0170	0.3527
	WILSON	−4.5029	−15.8153	2264.3243	−7550.6066	0.0290	0.2730
	NRTL	24.0611	16.9207	−7482.0249	−6082.7347	0.0403	0.6400
H_2O/1-pentanol	UNIQUAC	−2.7402	−9.6022	818.2867	3461.7929	0.0408	0.6558
	WILSON	−0.3626	−48.9978	−148.4263	−10,000.0000	0.0375	0.7109
	NRTL	27.4105	1.2408	−8543.0889	−417.7151	0.0803	1.0454
H_2O/1-hexanol	UNIQUAC	−2.0196	−0.6236	621.4692	70.4286	0.7994	0.9868
	WILSON	16.5498	−80.0000	−6504.2841	−10,000.0000	0.0351	0.4534
	NRTL	633.9445	62.9729	−147,929.7011 147,929.70110	−9545.9276	0.0308	0.7138
H_2O/1-heptanol	UNIQUAC	−27.0665	−27.3470	10,000.0000	10,000.0000	0.0289	0.7988
	WILSON	−27.4899	−80.0000	10,000.0000	−10,000.0000	0.0274	0.3687

The experimental results indicate that the investigated H_2O–long-chain alcohol system exhibits highly non-ideal behavior. During the correlation process, although the NRTL model demonstrated the best correlation accuracy within the experimental data range, the optimized binary interaction parameters exhibited abnormally large absolute values. This clearly reveals the inherent mathematical limitations of local composition models in describing such extremely non-ideal systems characterized by strong asymmetry and a tendency for liquid-phase splitting. These abnormal parameters likely result from over-parameterization, where the model forces a perfect fit to a limited dataset, leading to physically meaningless values. Consequently, the obtained model parameters should be used strictly for interpolation within the experimental data range. Extrapolation for predictions is strongly discouraged, as it would yield unreliable results.

As indicated by the binary correlated interaction parameters and root mean square deviations (RMSD) listed in the table, all three activity coefficient models can satisfactorily correlate the H_2O/EM system. The figure demonstrates that each model is capable of describing this thermodynamic system with reasonable accuracy. Moreover, 1-pentanol, 1-hexanol, and 1-heptanol are all able to break the azeotropic system of H_2O/EM. Data indicate that the mass fractions of H_2O in their azeotropes with water are approximately 56%, 68%, and 83%, respectively.

From the experimental data, the azeotropic temperature of H_2O and EM is 99.9 °C, and the boiling point of the new azeotrope formed by the azeotrope and H_2O should be lower than that of the original azeotrope, while the temperature difference between the new azeotrope formed by 1-hexanol and 1-heptanol and H_2O is too small compared with the original azeotropic temperature, which will cause separation difficulties. Secondly, the larger the proportion of H_2O in the azeotropic composition, the less azeotropic dose required to remove the H_2O from the system, that is, the less separation load, which is more conducive to reducing the energy consumption of the subsequent distillation column. It can be seen from Table 13 that the azeotropic composition of the three azeotropic agents accounts for a large proportion of H_2O. In summary, 1-pentanol is the best choice among the three azeotropes.

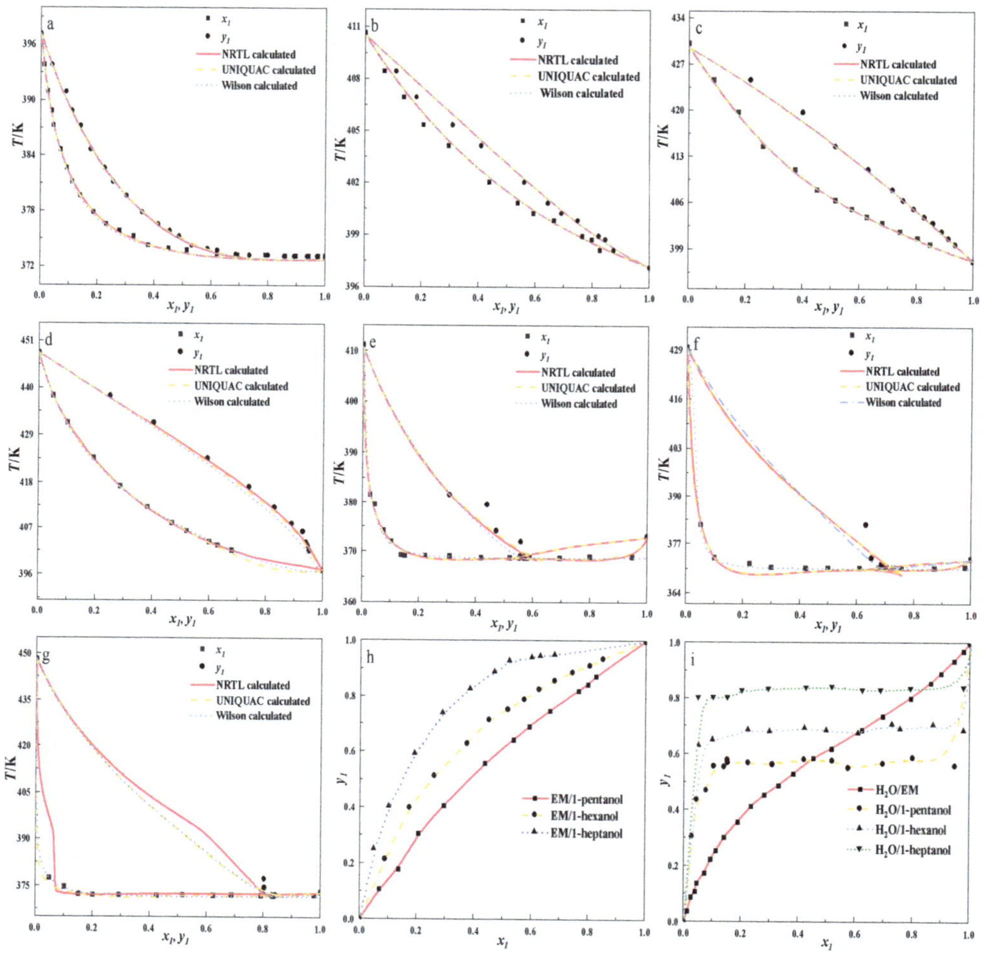

Figure 4. Experimental and calculated T-x_1-y_1 diagram. (**a**): H_2O/EM; (**b**): EM/1-pentanol; (**c**): EM/1-hexanol; (**d**): EM/1-heptanol; (**e**): H_2O/1-pentanol; (**f**): H_2O/1-hexanol; (**g**): H_2O/1-heptanol) Experimental y_1-x_1 diagram (**h**,**i**).

Table 13. Physical properties of azeotropes in H_2O and different substances (P = 101.325 kPa).

Azeotropic Substances	1-Pentanol	1-Hexanol	1-Heptanol
Boiling point/°C	138	157	176
Azeotropic temperature/°C	95.9	97.9	99.2
Azeotropic Composition (H_2O)/wt%	56	68	83

3.4. Molecular Interaction Simulation

Molecular interaction energy simulations depicted in Figure 5 enabled the calculation of ΔE via Equation (2), yielding values of −6.348 kJ/mol for H_2O/EM, 3.058 kJ/mol for EM/1-pentanol, 3.115 kJ/mol for EM/1-hexanol, 3.162 kJ/mol for EM/1-heptanol, −4.691 kJ/mol for H_2O/1-pentanol, −4.701 kJ/mol for H_2O/1-hexanol, and −4.694 kJ/mol for H_2O/1-heptanol. Systems exhibiting negative deviations (H_2O/EM, H_2O/1-pentanol, H_2O/1-hexanol, and H_2O-1-heptanol; all $\Delta E < 0$) corresponded to minimum H_2O activity coefficients of 0.9962, 0.6867, 0.7580. Conversely, systems showing positive deviations (EM/1-pentanol, EM/1-hexanol, and EM/1-heptanol; all $\Delta E > 0$) displayed maximum EM activity coefficients of 1.1857, 1.2116. Critically, these molecular simulation outcomes demonstrate quantitative consistency with experimental VLE data across all investigated systems.

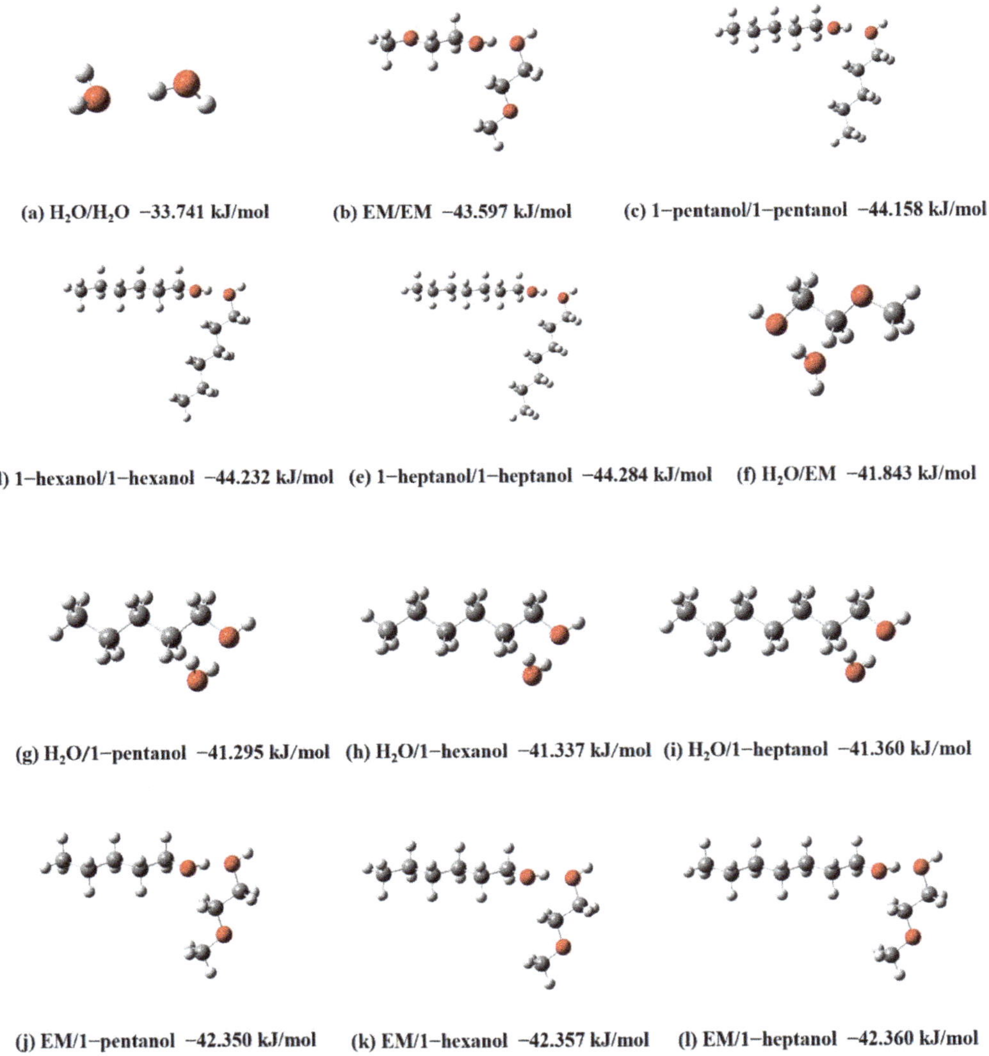

Figure 5. Molecular interaction simulation results of these components.

4. Separation of H_2O/EM via Azeotropic Distillation

4.1. Process Flow

To verify the feasibility of the selected entrainer for the separation, a process flow simulation was conducted. The wastewater from a pesticide plant contains 88 wt% H_2O, 10 wt% methyl cellosolve, and 2 wt% methanol. The simulated process flow is shown in Figure 6. With a feed flow rate of 1000 kg/h, the stream first enters T1, the light-ends removal column. The primary objective of T1 is to remove the light component (methanol) from the feedstock, with the distillate obtaining the light component at a mass fraction ≥ 0.999. The bottom mixture then proceeds to T2, the azeotropic distillation column (COL-MAIN). Here, under the influence of the entrainer 1-pentanol, H_2O is entrained and carried overhead. This overhead vapor is condensed and separated in a decanter (DECANTER). The upper organic phase, primarily consisting of n-pentanol, is returned to the azeotropic distillation column, with any lost entrainer being replenished. The lower aqueous phase enters the solvent recovery column (COL-REC). The bottom product from COL-REC is methyl cellosolve with a mass fraction ≥ 0.999. Simultaneously, the bottom stream of the solvent recovery column (COL-REC) produces H_2O with a mass fraction ≥ 0.999, while its distillate is an azeotropic mixture of H_2O and the entrainer. This distillate is subsequently recycled back to the azeotropic distillation column (COL-MAIN) for reuse.

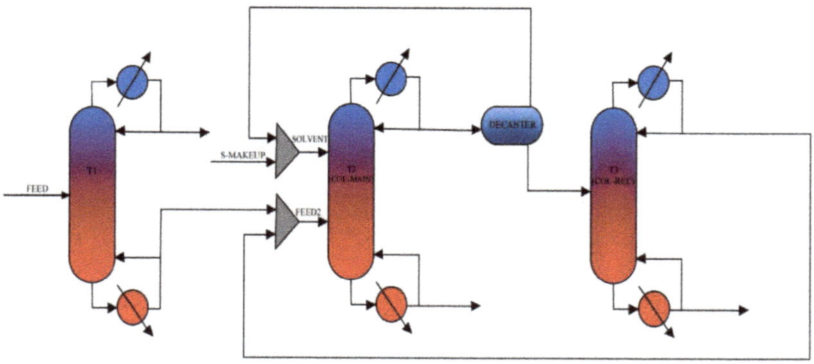

Figure 6. Process Flow Simulation.

4.2. Process Parameter Optimization

In the distillation process, key process parameters such as the feed flow rate of the entrainer, the number of theoretical stages, the feed stage location of the feedstock, and the reflux ratio have a crucial impact on the separation efficiency. Column T3, the solvent recovery column, is relatively straightforward to optimize; therefore, the primary focus was on optimizing columns T1 and T2. Sensitivity analysis on the process parameters related to columns T1 and T2 was conducted using Aspen Plus. The process was optimized by minimizing the estimated reboiler duty (Q), the thermal energy consumption. However, this represents only a preliminary stage of process evaluation. The results are shown in Figures 6 and 7.

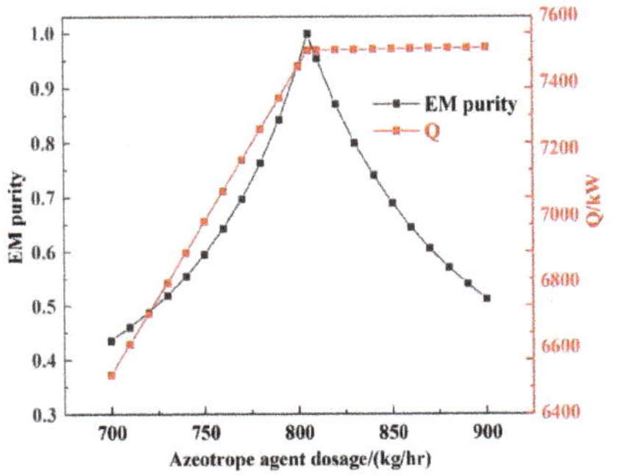

Figure 7. Effect of Azeotroping Agent dosage on Separation Efficiency.

As shown in Figure 7, the optimal dosage of the entrainer is 805.3 kg/h, at which point the separation target (i.e., an EM purity of 0.999) is achieved.

As can be seen from Figure 8, by integrating the optimization results of both the light-ends removal column T1 and the azeotropic distillation column T2 systems, the optimal process parameters for this azeotropic distillation system can be determined, as summarized in Table 14.

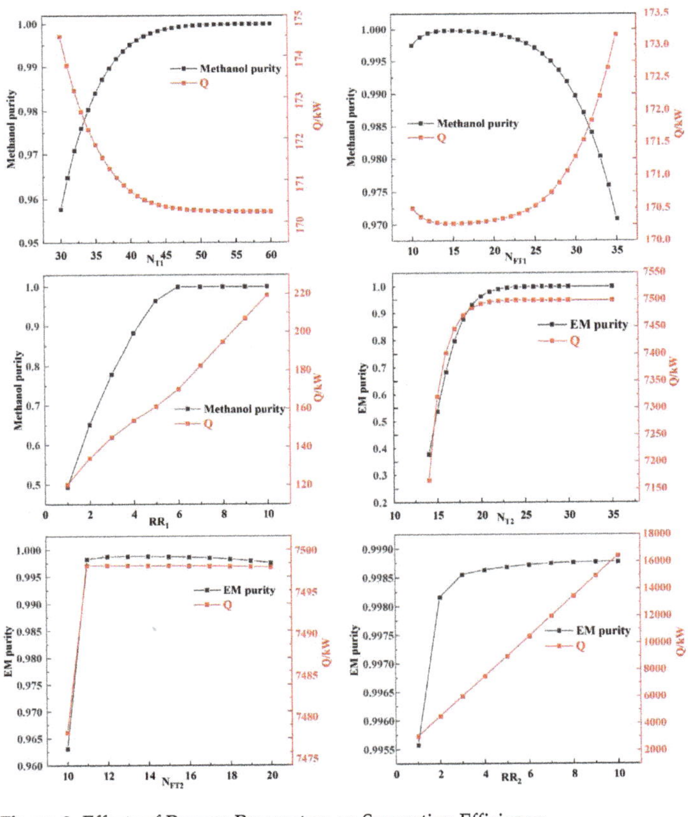

Figure 8. Effects of Process Parameters on Separation Efficiency.

Table 14. Results of Process Parameter Optimization.

Optimization Results	Light-Ends Removal Column T1	Azeotropic Distillation Column T2	Solvent Recovery Column T3
Number of Theoretical Stages	49	24	6
Feed Stage Location	14	13 (FEED2)	3
Reflux Ratio	6	4	1
Total Duty/kW	170.27	7498.98	105.65
Column Pressure/kPa	101.325	101.325	101.325
Entrainer Feed Rate/(kg/h)	/	805.3	/

5. Conclusions

VLE data for seven binary systems were measured at atmospheric pressure using a modified Othmer still, filling a significant data gap. All experimental data passed the thermodynamic consistency test via Herington's area method. Using the Aspen Plus process simulation software, the parameters for the NRTL, UNIQUAC, and Wilson activity coefficient models were regressed, yielding parameter sets applicable at atmospheric pressure. The results indicate that all three activity coefficient models can satisfactorily correlate the H_2O/EM system. Molecular interaction energy simulations agreed with the experimental VLE data, confirming the reliability of the data provided in this work. The analysis shows that 1-pentanol is the optimal choice among the three evaluated entrainers.

Subsequently, the azeotropic distillation process was simulated using Aspen Plus to evaluate the separation performance and identify the optimal operating parameters. This simulation provided the foundation for conceptual process design and preliminary thermodynamic assessment. Upon finalizing the conceptual design, addressing operational challenges such as emulsion formation and entrainer loss will become critical obstacles to overcome during subsequent detailed engineering design and pilot-scale scaling. These issues represent key focus areas for future work.

Author Contributions: Writing—review and editing, C.L.; writing—original draft preparation, J.Z.; software, J.R.; methodology, K.S.; validation, Y.S.; supervision, W.L.; project administration, J.L. All authors have read and agreed to the published version of the manuscript.

Funding: This work was financially supported by the National Natural Science Foundation of China (No. 22408082), the China Postdoctoral Science Foundation (No. 2023M740969), the Natural Science Foundation of Tianjin (No. 23JCQNJC00370), and the Tianjin Education Commission Scientific Research Project (No. 2023KJ293).

Data Availability Statement: The original contributions presented in this study are included in the article. Further inquiries can be directed to the corresponding author(s).

Conflicts of Interest: The authors declare no conflict of interest.

References

1. Carr, C.; Riddick, J.A. Physical Properties of Methanol-Water System. *Ind. Eng. Chem.* **1951**, *43*, 692–696. [CrossRef]
2. Vaghela, K.C.; Vankar, H.P.; Shah, K.N.; Rana, V.A. Dielectric and refractive index analysis of ethylene glycol monomethyl ether-methanol mixtures: Molecular interactions and machine learning based predictions. *J. Mol. Liq.* **2025**, *430*, 127756. [CrossRef]
3. Joshi, Y.S.; Kumbharkhane, A.C. Study of heterogeneous interaction in binary mixtures of 2-methoxyethanol-water using dielectric relaxation spectroscopy. *J. Mol. Liq.* **2011**, *161*, 120–124. [CrossRef]
4. Dong, Y.; Xu, F.; Xia, Y.; Yang, Q. Efficient separation of methanol, acetonitrile, and benzene ternary azeotropic mixture using ionic liquid in extractive distillation. *Sep. Purif. Technol.* **2025**, *360*, 131018. [CrossRef]

Gomes, J.P.; Silva, R.; Nunes, C.P.; Barbosa, D. An Alternative Green Solvent for 1,3-Butadiene Extraction. *Sustainability* **2025**, *17*, 3124. [CrossRef]

Waltermann, T.; Grueters, T.; Muenchrath, D.; Skiborowski, M. Efficient optimization-based design of energy-integrated azeotropic distillation processes. *Comput. Chem. Eng.* **2020**, *133*, 106676. [CrossRef]

Qi, J.; Li, Y.; Xue, J.; Qiao, R.; Zhang, Z.; Li, Q. Comparison of heterogeneous azeotropic distillation and energy-saving extractive distillation for separating the acetonitrile-water mixtures. *Sep. Purif. Technol.* **2020**, *238*, 116487. [CrossRef]

Torres Cantero, C.A.; Pérez Zúñiga, R.; Martínez García, M.; Ramos Cabral, S.; Calixto-Rodriguez, M.; Valdez Martínez, J.S.; Mena Enriquez, M.G.; Pérez Estrada, A.J.; Ortiz Torres, G.; Sorcia Vázquez, F.d.J.; et al. Design and Control Applied to an Extractive Distillation Column with Salt for the Production of Bioethanol. *Processes* **2022**, *10*, 1792. [CrossRef]

Bejarano, A.; de la Fuente, J.C. Isobaric (vapor +liquid) equilibria for the binary systems (1-methoxy-2-propanol+2-methoxyethanol), (2-butanone+2-methoxyethanol) and (water+2-methoxyethanol) at pressures of (74.5, 101.3, and 134.0) kPa. *J. Chem. Thermodyn.* **2015**, *87*, 88–95. [CrossRef]

10. Ochi, K.; Kojima, K. A measurement of vapor-liquid equilibria at extreme dilution. *J. Chem. Eng. Jpn.* **1987**, *20*, 6–10. [CrossRef]
11. Krishnaiah, A.; Reddy, K.V.R.; Devarajulu, T.; Ramakrishna, M. Isobaric vapour–liquid equilibria of methyl cellosolve–aliphatic alcohol systems. *Fluid Phase Equilibria* **1999**, *165*, 59–66. [CrossRef]
12. Havasi, D.; Mizsey, P.; Mika, L.T. Vapor–Liquid Equilibrium Study of the Gamma-Valerolactone–Water Binary System. *J. Chem. Eng. Data* **2016**, *61*, 1502–1508. [CrossRef]
13. Li, W.; Du, Y.; Li, J.; Chen, X.; Guo, S.; Zhang, T. Isobaric vapor-liquid equilibrium for acetone + methanol system containing different ionic liquids at 101.3 kPa. *Fluid Phase Equilibria* **2018**, *459*, 10–17. [CrossRef]
14. Korotkova, T.G. Parameters of the UNIQUAC Model for Describing the Vapor–Liquid Phase Equilibrium of the Isotope Mixtures of Hydrogen H2–D2, H2–HD, HD–D2. *Russ. J. Phys. Chem. A* **2024**, *98*, 92–105. [CrossRef]
15. Fenglian, L.; Lei, Y.; Muhammad, N. A new fractal transform for the approximate solution of DRINFELD–SOKOLOV–WILSON model with fractal derivatives. *Fractals* **2023**, *31*, 1–9.
16. Foute, A.J.; Ekosso, M.C.; Fotsin, H.; Fai, L.C. Effects of Gaussian thermal fluctuations on the thermodynamic of microtubules in Landau-Ginzburg-Wilson model. *Chin. J. Phys.* **2021**, *73*, 349–359. [CrossRef]
17. Cao, J.; Yu, G.; Chen, X.; Abdeltawab, A.A.; Al-Enizi, A.M. Determination of Vapor–Liquid Equilibrium of Methyl Acetate + Methanol + 1-Alkyl-3-methylimidazolium Dialkylphosphates at 101.3 kPa. *J. Chem. Eng. Data* **2017**, *62*, 816–824. [CrossRef]
18. Tang, G.; Ding, H.; Hou, J.; Xu, S. Isobaric vapor–liquid equilibrium for binary system of ethyl myristate + ethyl palmitate at 0.5, 1.0 and 1.5kPa. *Fluid Phase Equilibria* **2013**, *347*, 8–14. [CrossRef]
19. Liu, Y.; Liu, Y.; Wang, H. Vapor-liquid equilibrium and topological analysis of ethyl propionate synthesized by reaction distillation. *J. Chem. Thermodyn.* **2022**, *171*, 106810. [CrossRef]
20. Malloum, A.; Conradie, J. Molecular simulations of the adsorption of aniline from waste-water. *J. Mol. Graph. Model.* **2022**, *117*, 108287. [CrossRef]
21. Li, J.; Huang, Q.; Wang, G.; Wang, E.; Ju, S.; Qin, C. Experimental study of effect of slickwater fracturing on coal pore structure and methane adsorption. *Energy* **2022**, *239*, 122421. [CrossRef]
22. Bauer, G.; Gross, J. Phase Equilibria of Solid and Fluid Phases from Molecular Dynamics Simulations with Equilibrium and Nonequilibrium Free Energy Methods. *J. Chem. Theory Comput.* **2019**, *15*, 3778–3792. [CrossRef] [PubMed]
23. Sirirak, K.; Vao-soongnern, V. Molecular simulation for the effect of chain stiffness on polymer crystallization from the melts. *J. Mol. Liq.* **2023**, *387*, 122650. [CrossRef]
24. Sun, R.; He, H.; Yuan, L.; Wan, Y.; Sha, J.; Li, L.; Jiang, G.; Li, Y.; Li, T.; Ren, B. Thermodynamic analysis and molecular simulation of solid–liquid phase equilibrium of isoprenaline hydrochloride in eleven pure solvents at saturation. *J. Chem. Thermodyn.* **2021**, *160*, 106411. [CrossRef]
25. Wang, J.; Sun, J.; Shang, M.; Huang, C.; Rui, X. Isobaric vapour-liquid equilibrium for systems of propionic acid, butanoic acid and 1,4-butyrolactone at 15.0 kPa. *J. Chem. Thermodyn.* **2023**, *186*, 107143. [CrossRef]
26. Shi, K.; Zheng, H.; Sun, Y.; Wang, H.; Fang, J.; Li, C.; Liu, J. Vapor-liquid equilibrium for 1,4-butanediol, ethylene glycol and 3-methyl-1,5-pentanediol systems under different vacuum conditions. *Vacuum* **2024**, *230*, 113752. [CrossRef]
27. Zhai, X.; Wu, Y.; Tian, B.; Yao, Y.; Wu, B.; Chen, K.; Ji, L. Isobaric Vapor–Liquid Equilibrium for the Binary System of Cumene + Acetophenone at 101.3, 30.0, and 10.0 kPa. *J. Chem. Eng. Data* **2022**, *67*, 669–675. [CrossRef]
28. Han, X.; Wang, Y.; Wang, Y.; Zhao, H.; Li, Q. Vapor–Liquid Equilibrium Experiment and Process Simulation for Methanol and Ethyl Formate at 101.3 kPa. *J. Chem. Eng. Data* **2021**, *66*, 1929–1938. [CrossRef]
29. Rodrigues, V.H.S.; Almeida, R.N.; Vargas, R.M.F.; Cassel, E. Vapor pressure and vapor-liquid equilibrium data for eugenol/caryophyllene binary system at low pressures by experimental and predictive methods. *J. Chem. Thermodyn.* **2022**, *168*, 106725. [CrossRef]

30. Li, F.; Li, M.; Guo, J.; Hu, X.; Zhu, J.; Li, Q.; Zhao, H. Isobaric Vapor–Liquid Equilibrium Experiment of N-Propanol and N-Prop Acetate at 101.3 kPa. *J. Chem. Eng. Data* **2023**, *68*, 358–365. [CrossRef]
31. Shi, Y.; Wu, H.; Xu, J. Studies on Vapor–Liquid Equilibrium of Binary Systems Containing the MMA Component under Norm Atmospheric Pressure. *J. Chem. Eng. Data* **2022**, *67*, 3650–3660. [CrossRef]

separations

MDPI

Article

Efficient Extraction of 1,2-Dichloroethane from Wastewater Using Hydrophobic Deep Eutectic Solvents: A Green Approach

Irfan Wazeer, Abdullah Omair, Lahssen El Blidi, Salim Mokraoui, Emad Ali and Mohamed K. Hadj-Kali *

Chemical Engineering Department, College of Engineering, King Saud University, P.O. Box 800, Riyadh 11421, Saudi Arabia; 443106172@student.ksu.edu.sa (A.O.); lelblidi@ksu.edu.sa (L.E.B.); smokraoui@ksu.edu.sa (S.M.); amkamal@ksu.edu.sa (E.A.)

* Correspondence: mhadjkali@ksu.edu.sa

Abstract: This study provides a thorough examination of the utilization of hydrophobic deep eutectic solvents (HDESs) for the extraction of 1,2-dichloroethane (1,2-DCA) from effluent, with an emphasis on a sustainable and environmentally friendly approach. The extraction efficacy of six HDES systems was initially evaluated, and the combinations of thymol/camphor (Thy/Cam) and menthol/thymol (Men/Thy) exhibited superior performance. Subsequently, these two HDESs were chosen for a comprehensive parametric analysis. The impact of contact time demonstrated that extraction equilibrium was reached at 15 min for both systems, thereby achieving a balance between high efficiency and time efficiency. Next, the impact of the HDES-to-water mass ratio was investigated. A 1:1 ratio was determined to be the most effective, as it minimized solvent consumption and provided high efficiency. An additional examination of the molar ratios of the HDES components revealed that the 1:1 ratio exhibited the most effective extraction performance. This was due to the fact that imbalances in the solvent mixture resulted in diminished efficiency as a result of disrupted molecular interactions. The extraction efficiency was significantly influenced by the initial concentration of 1,2-DCA, with higher concentrations resulting in superior results as a result of the increased mass transfer driving forces. In general, the Men/Thy and Thy/Cam systems have shown noteworthy stability and efficiency under different conditions, which makes them highly suitable for large-scale applications.

Keywords: 1,2-dichloroethane; liquid–liquid extraction; hydrophobic DES; wastewater treatment

1. Introduction

The rapid growth of the world's population and rapid industrial revolution requires high water consumption, which has led to an increase in the amount of wastewater to be treated [1]. A major problem is the direct discharge of industrial wastewater into water bodies, which can cause serious problems to flora and fauna [2]. One of the major water pollutants is chlorinated aliphatic hydrocarbons (CAHs), including perchloroethylene (PCE), trichloroethylene (TCE), 1,1,1-trichloroethane (TCA), and 1,2-dichloroethane (1,2-DCA). Owing to their high density and poor water solubility, CAHs are persistent and potentially hazardous pollutants that form dense nonaqueous liquids in water [3]. Among CAHs, 1,2-DCA has become an attractive topic for researchers because it has not been extensively studied like other CAHs and because it is persistent, toxic, and potentially carcinogenic owing to its high demand for the synthesis of vinyl chloride monomers [3,4]. Hydrophobic deep eutectic solvents (HDESs) have recently attracted interest as environmentally friendly substitutes for the extraction of nonpolar contaminants from water. Terpene-based HDESs, derived from natural chemicals such as thymol and menthol, have favorable properties such as high hydrophobicity, low toxicity, and robust extraction efficacy.

The common use of 1,2-DCA is the production of vinyl chloride in chlorinated hydrocarbon reactions. The latter is the most important precursor in the production of polyvinyl chloride (PVC). Wastewater from PVC production is characterized by several pollutants, including 1,2-DCA [5]. The World Health Organization has set a permissible limit for 1,2-DCA in drinking water at 0.03 mg/L. The U.S. Environmental Protection Agency has also set a maximum level for 1,2-DCA in drinking water of 0.05 mg/L [6]. However, lower levels of 1,2-DCA in water are recommended.

1,2-DCA has attracted much attention in recent years because it is considered a hazardous substance for both the environment and human health. The properties of 1,2-DCA, such as high solubility, difficult adsorption, and stability, make it very difficult to deal with contaminated soil and groundwater [7]. The symmetrical structure with a single C-C bond, as well as the high energy for dissociation of the C-Cl bond and the high activation energy required for dechlorination, indicate that strong dissociation energies are required for the degradation of 1,2-DCA [3,7]. Therefore, it is important to find an effective and inexpensive method to remove 1,2-DCA from aquifers.

Researchers have attempted several methods to reduce or remove 1,2-DCA from wastewater. Lee et al. [8] investigated the adsorption of 1,2-DCA from water using activated carbon produced from beer lees. This study demonstrates the effectiveness of activated carbon in the removal of 1,2-DCA and highlights the potential of this low-cost adsorbent for water remediation. However, the strong attraction of water to 1,2-DCA interrupts adsorption. In addition, these physical treatments only transfer contaminants from the water to another phase and must be disposed of or further treated [9]. Another method is zero-valent iron or nano iron (nZVI), which has been used by many researchers to degrade 1,2-DCA. According to Baric et al. [10], this method is effective in treating chlorinated ethenes and ethanes (for example, PCE, TCE, and other chlorinated compounds) by degrading these pollutants into non-toxic compounds. In contrast, the degradation of 1,2-DCA has been shown to be resistant to reduction by nZVI. However, nZVI can be coupled with other compounds, such as dithionite, and it was found that coupled nZVI-dithionite was able to degrade >90% of 1,2-DCA over the course of a year. However, this method requires a long time to achieve a highly effective degradation [10,11]. Eydivand and Nikazar [12] used an advanced oxidation method to remove 1,2-DCA from simulated wastewater. They investigated different oxidation methods, including ultraviolet (UV) irradiation, H2O2/UV, and photocatalytic methods (using TiO2 and ZnO as photocatalysts), to degrade 1,2-DCA in a simulated synthetic solution to optimize the method. The removal of 1,2-DCA reached 90% using the TiO2/UV method. Jeong et al. [3] utilized Fenton oxidation (a powerful advanced oxidation process that generates hydroxyl radicals) to degrade 1,2-DCA. They investigated several reagents (dithionite, hydrosulfide, sulfite, persulfate (PS), sulfate radicals, and hydroxyl radicals), of which the hydroxyl radicals generated by the Fenton reaction were the most suitable oxidizing agents and degraded 92% of 1,2-DCA. Although this method showed good results in the removal of 1,2 DCA, it has some disadvantages, such as the toxicity of hydrogen peroxide and hydroxyl radicals, the need for acidic pH, the production of iron sludge, and the need for a high chemical input [13,14]. Liu et al. [15] investigated the degradation of 1,2-DCA with advanced reduction processes (ARPs). In their study, 1,2-DCA was successfully degraded with various ARPs by combining UV irradiation with different reagents (dithionite, sulfite, sulfide, and iron). More than 90% of the initial 1,2-DCA was removed within 20 min under alkaline conditions (pH 8.2, 9.0, and 11.0), while it took 130 min to achieve the same removal at pH 7.0. However, despite these positive results, there are no reports on the large-scale use of this technology [16]. Membrane filtration is a method used to remove contaminants from water. Yazdanbakhsh et al. [5] attained a very high removal of 1,2-DCA (96.7) from water using a nanofiltration membrane cell followed by microfiltration.

Biodegradation techniques have also been used to remove 1,2-DCA. According to some studies, 1,2-DCA can act as both an electron donor and an acceptor during aerobic and anaerobic reductive dechlorination, respectively [7]. Marines et al. [17] investigated the feasibility of chlorinated solvent removal by biostimulated and bioaugmented biological processes in permeable reactive barriers at a laboratory scale under anaerobic and aerobic conditions. They obtained 100% removal of 1,2-DCA during the anaerobic process, whereas the aerobic process showed a slightly lower 1,2-DCA removal efficiency of 98% after 75 days of testing. Despite the high biodegradation efficiency and effective dechlorination of 1,2-DCA to non-toxic ethylene [18], numerous variables, such as temperature, pH, and nutritional status, affect the biodegradation process. These variables are difficult to manage in real-world situations and change throughout the season. In addition, the biodegradation process is time-consuming, and the resulting byproducts are not always desired [7]. Table 1 shows the comparison of various methods used for the removal of CAHs. The table signifies the strength of various approaches based on key performance indicators such as distribution coefficients (D), selectivity (S), absorption rate (AR), and absorption capacity (AC).

Table 1. Comparison of various methods used for the removal of CAHs.

Material	CAH	Media	Process	Conditions	Efficiency	Test	Ref.
PDS [1]	PCE	Groundwater	Oxidation	T = 25 °C, Time = 12 min	High removal efficiency	GC	[19]
PDS	PCE	water	Electro-Fenton	PH = 5, PCE conc = 15 mg/L	98.1%	NA	[20]
Ni-nZVI	TCE	Water	nZVI reduction	5 wt% Ni, T = 26 °C, pH = 7–8.8	100%	GC-MSD	[21]
PS with Menthol	CTC [2]	Water	PS activation	Time = 48 h	~100%	GC/MS	[22]
PS	TCA	Water	Sono-activated oxidation	T = 20 °C, pH = 7, PS:TCA = 10:1	~100%	GC/MS	[23]
TiO_2	DCM [3]	Waste Gas	Advanced oxidation	T = 20 °C, PH = 7	95.5%	NA	[24]
[Bmim][DCA] [Bmim][OAc] [Bmim][SCN] [Bmim][PF_6] [Bmim][BF_4]	1,2-DCA	NA	Absorption	T = 313.15 K, P = 1 atm	0.37 g EDC/g IL 0.33 g EDC/g IL 0.30 g EDC/g IL 0.21 g EDC/g IL 0.24 g EDC/g IL	NA	[25]
[EMIM][BF_4] [EMIM][BF_4] [EMIM][TF_2N] [EMIM][TF_2N]	1,2-DCA TCEt [4] 1,2-DCA TCEt	NA	Absorption	T = 298 K, P = 1 atm, flow rate = 15 mL/min	AR = 95.8% AR = 84.9% AR = 94.9% AR = 82.9%	GC GC GC GC	[26]
[EMIM][Ac]	DCM	NA	Absorption	T = 303 K	AC = 756.55 (mg/g)	GC	[27]

Table 1. *Cont.*

Material	CAH	Media	Process	Conditions	Efficiency	Test	Ref
Sulfolane	1,2-DCA	Cyclo-hexane	Liquid–liquid extraction	T = 313.15 K, P = 1 atm	D = 1.83–2.08, S = 23.50–47.69	GC	[28]
DMSO [5]					D = 2.43–3.18, S = 12.31–39.32		
DMF [6]					D = 1.46–2.11, S = 2.11–4.28		
1,4-butanediol					D = 0.44–0.54, S = 9.16–14.44		

[1] Polydimethylsiloxane, [2] carbon tetrachloride, [3] dichloromethane, [4] 1,1,2-trichloroethane, [5] dimethyl sulfoxide, [6] N,N-dimethylmethanamide.

Liquid–liquid extraction (LLE) was utilized effectively in this study to remove 1,2-DCA from polluted water. This technique is user-friendly, can handle thermally unstable and high-boiling substances with ease, and is known for its high efficiency and selectivity. Additionally, it is economically dependable and can be seamlessly scaled up. Therefore, this method is widely used at the laboratory scale and in industrial processes [29–31]. LLE involves the extraction of a solute or an analyte from a nonpolar solvent and an aqueous polar solution. It utilizes the principle of differential solubility and partition equilibrium to extract the desired material from two different immiscible solvents [32].

Various solvents, such as hydrocarbons, esters, alcohols, halogens, and ketones, can be used in LLE applications. Although they have good properties (readily available, high solvent power), they have some disadvantages, such as high volatility, high cost, and toxicity [31]. However, there are environmentally friendly alternatives, such as ionic liquids (ILs) and deep eutectic solvents (DESs) [33]. Both have interesting properties, such as a low melting point, low vapor pressure, non-flammability, and biodegradability [34]. In addition, DESs have advantages over ILs because they are cheaper, can usually be synthesized easily, are less toxic, and are often biodegradable, resulting in DESs receiving more attention from researchers [35]. In 2015, a new subclass of DESs with hydrophobic properties was discovered as an effective method for extracting nonpolar organic and inorganic molecules from aqueous solutions [36]. The first HDES was introduced by van Osch et al. [37] and consists of various quaternary ammonium salts with decanoic acid. HDESs are formed by mixing two or more (usually hydrophobic) components, resulting in a eutectic composite with a melting point lower than that of one of its components.

HDESs represent a novel class of green solvents specifically designed for the efficient extraction of non-polar pollutants from aqueous environments. Their low water solubility, environmental compatibility, and ease of synthesis make them attractive alternatives to conventional solvents and ionic liquids. In this class, terpene-based HDESs, which consist of natural compounds such as thymol, menthol, and camphor, have particularly promising properties due to their strong hydrophobic interactions, low viscosity, and biodegradability. These properties improve mass transfer during extraction and minimize the environmental and safety concerns associated with conventional solvents. In this study, both terpenes and other hydrophobic HDESs were evaluated for their efficacy in the removal of 1,2-DCA, a persistent and toxic pollutant.

The aim of this study is to develop an efficient method for the removal of 1,2-DCA from aqueous media using HDESs. First, six HDESs were investigated for the extraction of 1,2-DEA. All the HDESs were prepared at a molar ratio of 1:1. After the initial screening, the best HDES with the highest extraction efficiency was selected for further investigation. This study aimed to optimize the use of HDESs by investigating the effects of various parameters, including the contact time, molar ratio of the HDES, mass ratio of HDES to aqueous solution, and initial concentration of 1,2-DCA. A thorough investigation of these parameters will improve our understanding of the extraction process and ensure its effectiveness and sustainability. This approach aims to provide an environmentally friendly and effective solution for the removal of 1,2-DCA from contaminated water.

2. Materials and Methods

Table 2 shows the list of chemicals used in this experiment, along with their molecular weights, CAS No., sources, and mass fractions. The chemicals were used as received without further purification. Naphthalene was used as an internal standard material.

Table 2. Name, CAS No., supplier, and purity of the chemicals.

Chemicals	Cas Number	Source	MW (g/mole)	Purity *
1,2-DCA	107-06-2	Thermo Scientific (New Jersey, USA)	98.96	99.8
Naphthalene	91-20-3	Sigma-Aldrich (Schnelldorf, Germany)	128.17	≥0.99
Menthol	89-78-1	Sigma-Aldrich (St. Louis, MO, USA)	156.27	≥0.95
Thymol	89-83-8	BDH Laboratory (West Yorkshire, UK)	150.22	≥0.99
Decanoic acid	334-48-5	Sigma-Aldrich (Petaling Jaya, Malaysia)	172.26	≥0.98
Coumarin	91-64-5	Thermo Scientific (Les Ulis, France)	146.14	≥0.99
Camphor	76-22-2	Acros Organics (Shanghai, China)	152.23	≥0.96
Hydrocinnamic acid	501-52-0	Sigma-Aldrich (St. Louis, MO, USA)	150.17	≥0.99
Naphthalene	91-20-3	Sigma-Aldrich (Schnelldorf, Germany)	128.17	≥0.99
Toluene	108-88-3	Scarlau (Sentmenat, Spain)	92.14	≥0.99
Ethanol	64-17-5	Sigma-Aldrich (Schnelldorf, Germany)	46.07	≥0.99

* The purity (in mass fraction) of the chemicals was provided by the supplier.

2.1. Preparation of HDESs

The HDESs were prepared by mixing hydrogen bond donor (HBD) and hydrogen bond acceptor (HBA). Coumarin and camphor can act as HBAs, while menthol, thymol, decanoic acid, and hydrocinnamic acid can act as both HBAs and HBDs. The HDESs were prepared by mixing and stirring the components at a constant speed of 350 rpm and a temperature of 80 °C under magnetic stirring until a homogeneous liquid was formed [38,39]. All HDESs were prepared in a 1:1 molar ratio. Optimized molecular structure of the individual components of the HDESs are shown in Figure 1.

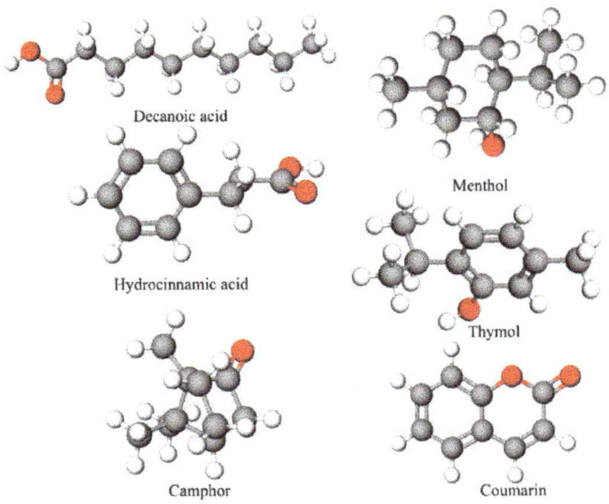

Figure 1. Optimized molecular structure of the individual components of the HDESs.

2.2. Experimental Procedure

A liquid–liquid extraction process was implemented in this investigation to effectively eliminate 1,2-DCA from water through the utilization of HDESs. A schematic of the experimental procedure is shown in Figure 2. Initially, a stock solution of 2.5 g of 1,2-DCA per 100 mL of deionized water was prepared for the screening of six distinct HDESs. This solution was treated with a Millipore Q-water system. The Eppendorf Thermomixer was utilized to mix the feed solution with the HDES at a 1:1 mass ratio during the extraction procedure. The selected mass ratio provides an optimal balance between solvent consumption and extraction efficiency. Excessive use of HDESs may slightly improve extraction but increases costs and environmental impact. Conversely, too little solvent limits mass transfer and phase separation. Previous studies have shown that the 1:1 ratio is effective for HDES–water systems due to favorable dispersion and reduced emulsion formation [33,34]. The mixer was operated at 2000 rpm and 25 °C for 30 min. A speed of 2000 rpm was chosen to ensure sufficient shear and mixing intensity to overcome interfacial resistance and allow rapid mass transfer, while avoiding the formation of emulsions that could hinder separation [33]. Extraction was performed at 25 °C to match standard environmental conditions and maintain consistency with previous HDES-based extraction studies. This temperature reduces volatility-related losses of 1,2-DCA and maintains the structural integrity of the terpene-based solvents. Following the mixing process, the samples were centrifuged at 5000 rpm for 10 min to facilitate phase separation, which resulted in the formation of two distinct layers. This speed was sufficient to overcome the differences in density and viscosity between the HDES and water and to ensure complete stratification into two separate layers.

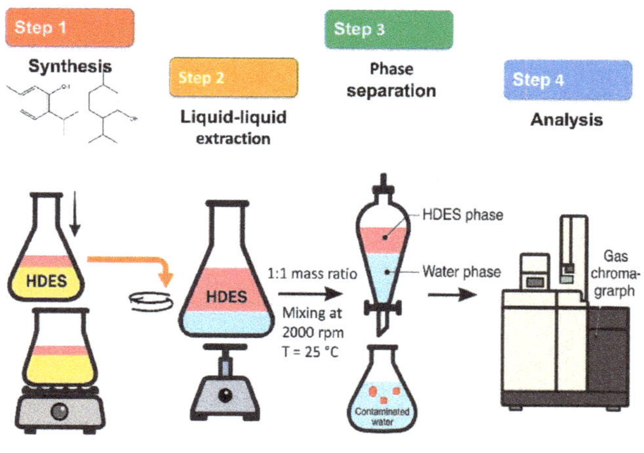

Figure 2. Schematic representation of the experimental procedure.

The HDES-phase samples were diluted with toluene, while the water-phase samples were diluted with toluene and ethanol. Samples from both the HDES and the water phases were collected. Following this, the samples were subjected to gas chromatography (GC) analysis, with naphthalene serving as an internal standard due to its thermal stability within the 300–400 °C range [40]. The results were validated by measuring average uncertainties, and each measurement was conducted in triplicate to assure accuracy.

The extraction efficiency (E %) of 1,2-DCA was evaluated using Equation (1):

$$E\,\% = \left(\frac{c_{1,2-DCA}^{initial} - c_{1,2-DCA}^{final}}{c_{1,2-DCA}^{initial}}\right) \times 100\% \quad (1)$$

where $c_{1,2-DCA}^{initial}$ is the initial 1,2-DCA concentration in the raffinate phase and $c_{1,2-DCA}^{final}$ is the final 1,2-DCA concentration in the raffinate phase.

GC was used to quantitatively analyze the presence of 1,2-DCA in both the HDES and the water phases. The study was performed with a Shimadzu GC-2010 Pro instrument, which was outfitted with a flame ionization detector and an HP-5 column under the operating conditions listed in Table 3. The HP-5 column is composed of 5% diphenyl and 95% dimethylpolysiloxane. It has a length of 30 m, an inner diameter of 0.32 mm, and a film thickness of 0.25 µm. Helium was used as the carrier gas in split mode.

Table 3. Gas chromatography operating conditions.

GC Parameter	Value
Temperature of detector (°C)	320.0
Temperature of injector (°C)	320.0
Helium carrier gas total flow (mL/min)	40.0
Oven program	40 °C for 2 min 180 °C to 300 °C Hold time: 0 min Rate: 30 °C/min

In order to avoid contamination and maintain the accuracy of the data, the GC liner was thoroughly cleansed following each analysis, specifically to eliminate any remaining HDES components. Naphthalene was chosen as the internal standard because of its high temperature stability and minimal interaction with 1,2-DCA, which makes it well-suited for precise calibration and quantification. Reliable measurements were aided by the use of naphthalene, which was free from interference in the target analyte matrix. The calibration curves were generated by comparing the response of naphthalene to 1,2-DCA. These curves showed a high level of linearity, with correlation coefficients of more than 0.99. This guaranteed accurate measurement of 1,2-DCA concentrations in the samples. Figure 3 displays the GC calibration curve for 1,2-DCA and naphthalene. Karl Fischer titration (Aquamax Karl-Fischer, GR Scientific Ltd., Halle, Germany) was used to quantify the water content of pure HDES and after extraction. Briefly, around 100 mg of each HDES was weighted and injected into the Karl-Fisher titrator cell. The titration was repeated three times, and an average value was calculated.

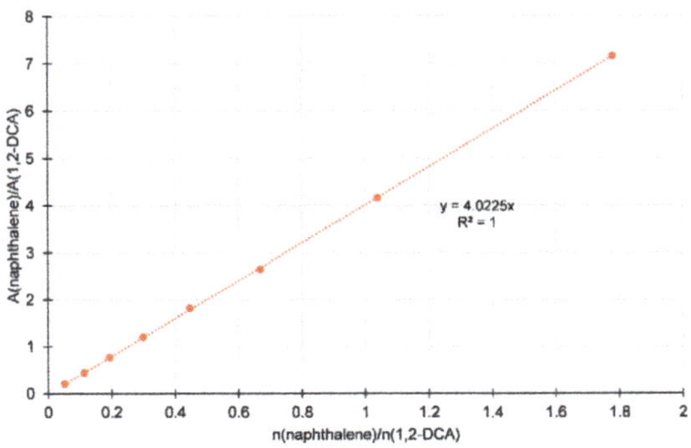

Figure 3. GC calibration curve of 1,2-DCA and naphthalene.

2.3. Selection of HDESs

The selection of HDESs for this study was conducted in a systematic manner to guarantee their efficacy in extracting 1,2-DCA from effluent. These particular HDESs were selected due to their known hydrophobicity and their capacity to generate stable eutectic mixtures at room temperature, a critical factor in the efficient separation of phases. Terpenes, including thymol and menthol, were chosen for their well-documented ability to produce stable HDESs. This property is further enhanced when combined with carboxylic acids, such as decanoic acid [41,42]. The selection process was influenced by the previous literature, which emphasized the advantages of utilizing natural, biodegradable components that are consistent with the principles of green chemistry. The hydrophobic nature of terpene-based components, such as thymol and menthol, facilitates the efficient separation of phases from the aqueous solution, while the reduced viscosity they provide enhances mass transfer during the extraction process [43]. Table 4 displays the list of HDESs that were evaluated in this study.

Table 4. List of prepared HDESs and their water content.

HDES	Component 1	Component 2	Mole Ratio	Water Content (%)
Men/Thy	Menthol	Thymol	1:1	0.011
Men/Dec	Menthol	Decanoic acid	1:1	0.213
Thy/Dec	Thymol	Decanoic acid	1:1	0.018
Hyd/Dec	Hydrocinnamic acid	Decanoic acid	1:1	0.049
Thy/Cou	Thymol	Coumarin	1:1	0.017
Thy/Cam	Thymol	Camphor	1:1	0.079

3. Results and Discussion

3.1. Cross-Contamination of HDESs with Water

When evaluating the suitability of HDESs for wastewater treatment applications, it is crucial to consider the potential for water contamination. The study focused on examining the interaction between HDESs and water during the LLE process. The HDESs were combined with water in a 50% v/v ratio and stirred for 2 h to ensure complete interaction between the two phases. Afterwards, there was a period of settling to naturally separate the phases. After completing this procedure, the HDES-phase samples were carefully examined to assess the level of water contamination. Figure 4 illustrates that the majority of HDESs exhibited minimal water absorption, suggesting excellent hydrophobicity and stability throughout the extraction process. Thy/Dec exhibited a significantly greater water absorption level in comparison to the other samples, with a water content of around 3.6%. It appears that while this HDES may have impressive extraction capabilities, its higher water solubility could potentially affect its separation efficiency and long-term stability in industrial applications. The other HDESs displayed water content below 3%, confirming their classification as neutral solvents with low water solubility.

This research aligns with previous studies on terpene-based HDESs, which typically show minimal interaction with water [41,44–46]. It is evident from the results that most of the tested HDESs exhibited a remarkable ability to retain their hydrophobic properties, even when exposed to water for extended periods of time. This feature is especially beneficial for processes that involve frequent reuse of solvents. It helps to minimize the loss of solvents and ensures consistent extraction efficiency over time.

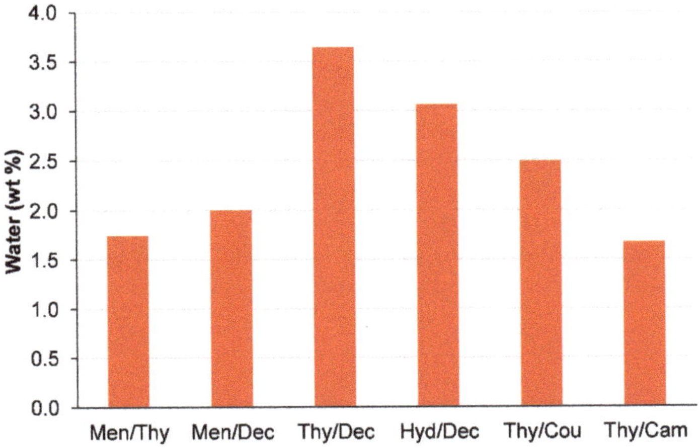

Figure 4. Water solubility of the prepared HDESs. HDES-to-water volume ratio = 1:1; stirring at 350 rpm for 2 h at 25 °C.

3.2. Screening of HDESs

During the initial screening phase, we evaluated six HDESs systems to determine their effectiveness in extracting 1,2-DCA from wastewater. The combinations that were tested include Men/Thy, Men/Dec, Thy/Dec, Hyd/Dec, Thy/Cou, and Thy/Cam. The extraction efficiencies showed variation, with Men/Thy and Thy/Cam demonstrating the highest performance, achieving efficiencies of 86.42% and 87.32%, respectively (Figure 5). These results highlight the superior extraction capabilities of these two HDES systems, making them prime candidates for further study.

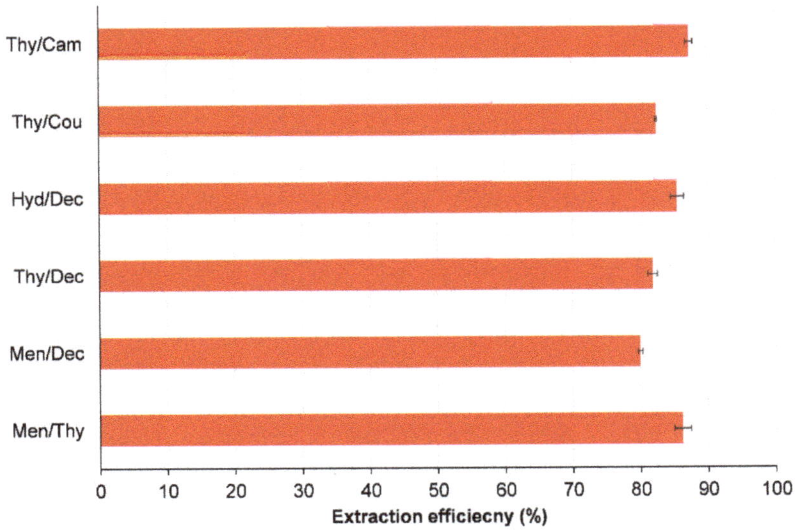

Figure 5. Extraction efficiency (%) of 1,2-DCA by various HDESs. Initial concentration of 1,2-DCA is 5000 ppm; HDES-to-water mass ratio = 1:1; vortex mixing, 2000 rpm for 30 min at 298.15 K; centrifugation at 5000 rpm for 10 min.

The improved performance of Men/Thy and Thy/Cam can be ascribed to the distinctive molecular interactions present in these solvent pairs. The potential synergistic effect of combining menthol and thymol is believed to be due to the unique properties of each compound. Menthol, known for its flexibility, and thymol, with its aromatic structure, are believed to generate an optimal microenvironment that enhances the extraction of 1,2-DCA. The Thy/Cam system leverages the aromatic interactions of thymol and the strong hydrophobic character of camphor to enhance its ability to solubilize and extract the target contaminant. Based on the encouraging preliminary findings, the Men/Thy and Thy/Cam configurations were chosen for a more comprehensive parametric analysis. This analysis aims to enhance their performance by optimizing various parameters across different conditions.

3.3. *Eeffect of Contact Time*

To enhance the efficiency of 1,2-DCA extraction, the influence of contact time was assessed using the top two HDESs from the screening phase, namely Thy/Men and Thy/Cam. For each system, extraction efficiency was determined across a range of contact times, spanning from 5 to 60 min. The results for the Thy/Men system demonstrated a steady increase in efficiency from 85.51% at 5 min to a peak of 86.64% at 45 min. However, any further improvement beyond 15 min was limited, with efficiency stabilizing around 86%, suggesting that the extraction process achieved equilibrium relatively quickly, as can be seen in Figure 6.

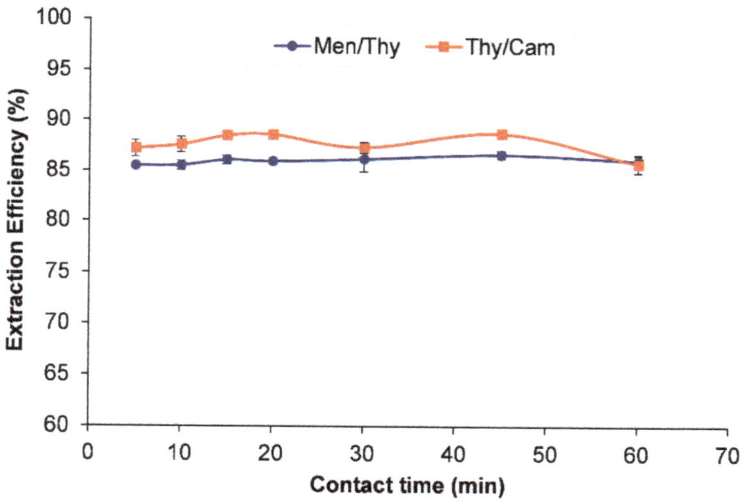

Figure 6. Effect of contact time on the extraction of 1,2-DCA using Men/Thy and Thy/Cam HDESs. Initial concentration of 1,2-DCA is 5000 ppm; HDES-to-water mass ratio = 1:1; vortex mixing, 2000 rpm for 30 min at 298.15 K; centrifugation at 5000 rpm for 10 min.

In contrast, the Thy/Cam system exhibited a more noticeable rise in extraction efficiency, reaching a maximum of 88.69% at 45 min before slightly decreasing to 85.77% at 60 min. The highest efficiency (88.57%) was attained at just 15 min, after which only slight gains were observed with extended contact times. Considering that both systems showed minimal improvement beyond 15 min while maintaining high extraction efficiency, this contact time was deemed optimal for further studies. Choosing 15 min balances the extraction performance with time efficiency, making the process effective and practical for real-world applications.

3.4. Eeffect of HDES-to-Water Mass Ratio

The ratio of HDES to water is essential in determining the efficiency of 1,2-dichloroethane extraction. In this study, we examined various mass ratios for the HDES systems Thy/Men and Thy/Cam in order to determine the most effective ratio for achieving maximum extraction. The results are depicted in Figure 7. For the Thy/Men system, the 1:1 mass ratio resulted in an efficiency of 86.42%, which showed a slight improvement when the HDES concentration was increased to a 2:1 and 3:1 ratio, reaching up to 87.26%. Nevertheless, the efficiency experienced a significant drop when the HDES concentration was reduced to ratios of 1:2, 1:3, and 1:4. This suggests that a decrease in HDES content has a detrimental impact on the solvent's ability to extract 1,2-dichloroethane. Although there was only a slight improvement at higher ratios, the 1:1 ratio was selected as the optimal choice due to its high efficiency and minimal solvent usage.

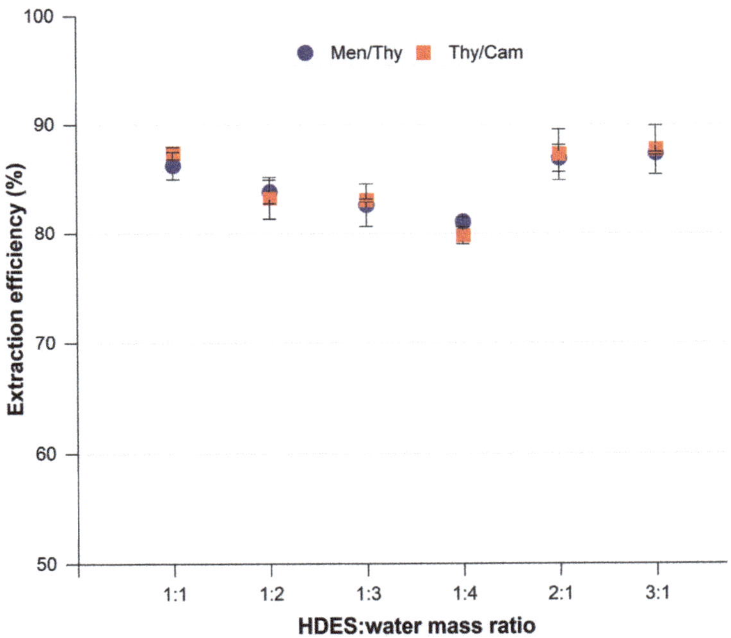

Figure 7. Effect of water-to-HDES mass ratio on the extraction of 1,2-DCA. Initial concentration of 1,2-DCA is 5000 ppm; vortex mixing, 2000 rpm for 15 min at 298.15 K; centrifugation at 5000 rpm for 10 min.

For the Thy/Cam system, it is worth noting that the 1:1 ratio demonstrated an efficiency of 87.32%. By increasing the HDES concentration to ratios of 2:1 and 3:1, there were only slight improvements observed, with the highest efficiencies reaching 87.61%. On the other hand, reducing the HDES content resulted in a significant decrease in extraction performance. The efficiency dropped to as low as 79.78% when using the 1:4 ratio. After careful evaluation of performance and solvent economy, the optimal condition for this system was determined to be a 1:1 ratio. This ratio ensures a high extraction efficiency while minimizing the excessive use of HDESs, making it a practical choice for further parametric studies and real-world applications.

3.5. Eeffect of HDES Molar Ratio

The extraction effectiveness of hydrophobic deep eutectic solvents (HDESs) is usually influenced by the molar ratio of their components, since it affects the interactions between the solvent components and the contaminant being targeted. The Thy/Men system had the maximum extraction efficiency of 86.42% while using a 1:1 molar ratio. The appropriate chemical interactions between menthol and thymol in this balanced ratio enabled the solvent combination to effectively remove 1,2-dichloroethane. The extraction efficiency decreased when the ratio was altered to have an excess of either menthol (1:2 and 1:3) or thymol (2:1 and 3:1), as reported in Figure 8. The decrease in solubility is probably caused by an uneven distribution of the solvent molecules, where an excess of one component hinders the formation of optimal hydrogen bonding and van der Waals interactions necessary for efficient dissolution of the contaminant.

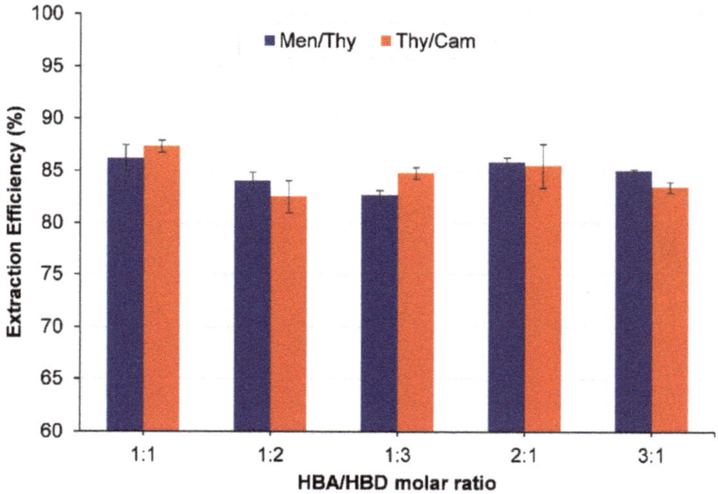

Figure 8. Effect of HDES molar ratio on the extraction of 1,2-DCA. Initial concentration of 1,2-DCA is 5000 ppm; HDES-to-water mass ratio = 1:1; vortex mixing, 2000 rpm for 15 min at 298.15 K; centrifugation at 5000 rpm for 10 min.

In the Thy/Cam system, similar behavior was observed, with the highest efficiency attained at a 1:1 molar ratio, resulting in an extraction efficiency of 87.32%. The efficiency decreased significantly when the ratio was adjusted to favor either thymol or camphor. For example, when the molar ratio was 1:2 (with a preference for camphor), the efficiency decreased to 82.37%, and at a ratio of 3:1 (with a preference for thymol), it decreased to 83.43%. The decrease in efficiency resulting from modified ratios can be ascribed to the disturbance of the ideal solvent environment. An excessive quantity of either component might result in a dilution effect or an imbalance in the eutectic mixture, which diminishes the HDES's capacity to generate stable microenvironments for extracting the contaminant. The findings highlight the need to maintain a balanced 1:1 molar ratio in order to preserve the structural integrity and extraction capabilities of the HDESs.

3.6. Effect of Initial Concentration of 1,2-DCA

The efficiency of the HDES extraction process for 1,2-dichloroethane (1,2-DCA) is strongly influenced by the initial concentration of the contaminant in the aqueous phase. In the Thy/Men system, the extraction efficiency increased as the 1,2-DCA concentration rose from 2000 ppm to 7000 ppm. At lower concentrations, the efficiency was only 50.13%, but it improved significantly at higher concentrations, reaching 86.47% at 5000 ppm and 91.82% at 7000 ppm (see Figure 9). This suggests that the HDES becomes more effective at extracting the contaminant as its concentration increases, likely due to the increased driving force for mass transfer.

A similar trend was observed in the Thy/Cam system. Efficiency improved from 55.92% at 2000 ppm to 90.86% at 7000 ppm. At concentrations below 4000 ppm, the extraction performance was moderate, but it increased sharply once the concentration reached 5000 ppm and above, where efficiencies exceeded 87%. This indicates that the HDES systems can effectively capture more 1,2-DCA when it is present in greater quantities. This behavior is likely due to the enhanced partitioning of 1,2-DCA into the HDES phase as its concentration rises, leading to more efficient extraction. These findings suggest that higher initial concentrations of 1,2-DCA favor the extraction process, making these HDESs particularly suitable for applications involving high contaminant loads.

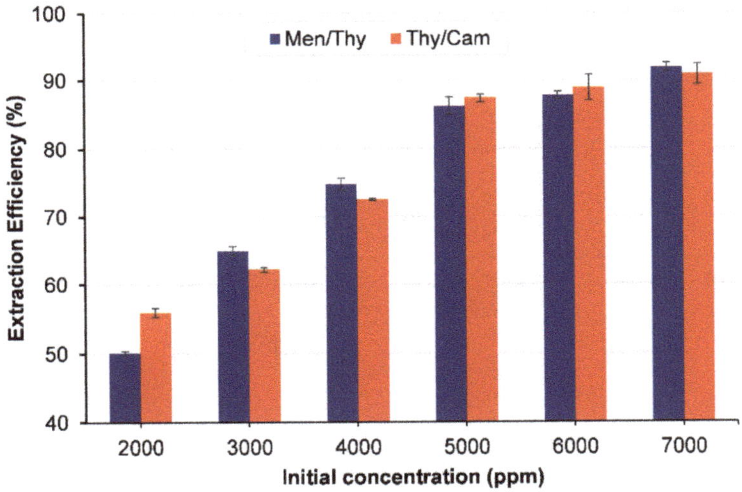

Figure 9. Effect of initial concentration of 1,2-DCA on the extraction efficiency. Vortex mixing, 2000 rpm for 15 min at 298.15 K; centrifugation at 5000 rpm for 10 min.

3.7. Extraction Mechanism

COSMO-RS (Conductor-like Screening Model for Real Solvents) is a quantum chemistry-based model used to predict the thermodynamic properties of liquids, especially liquid–liquid equilibria. It integrates quantum chemical calculations with statistical thermodynamics and provides insights into molecular interactions and solubility behavior without the need for experimental data [47]. In the context of COSMO-RS, two important results are the sigma profile and the sigma potential.

The sigma profile represents the distribution of surface screening charge densities (σ) across a molecule. Peaks in this profile indicate the presence of electron-rich (negative σ) or electron-poor (positive σ) regions and thus reflect hydrogen bond acceptor (HBA) or donor (HBD) sites. On the other hand, the sigma potential curve provides information about the chemical potential of a molecule in an environment with different surface polarity and thus characterizes the affinity of the molecule for different types of interactions, such as polar or non-polar [48]. Both profiles can be divided into three regions: positive polarity ($\sigma < -0.0084$ eÅ^2), non-polar (-0.0084 eÅ$^2 \leq \sigma \leq 0.0084$ eÅ^2), and negative polarity ($\sigma > 0.0084$ eÅ^2). The boundaries of the individual regions are highlighted by the dashed lines. The interactions between the molecules of a system are taken into account by comparing the peak area between the curves at negative and positive σ values.

In this study, the COSMO-RS model was used to investigate the extraction mechanism of 1,2-dichloroethane (1,2-DCA) from aqueous media using two hydrophobic deep eutectic solvents (HDESs): thymol/camphor (Thy/Cam) and menthol/thymol (Men/Thy). Figure 10 shows the sigma profiles of 1,2-DCA, water, and the two DESs, while Figure 11 shows their respective sigma potentials.

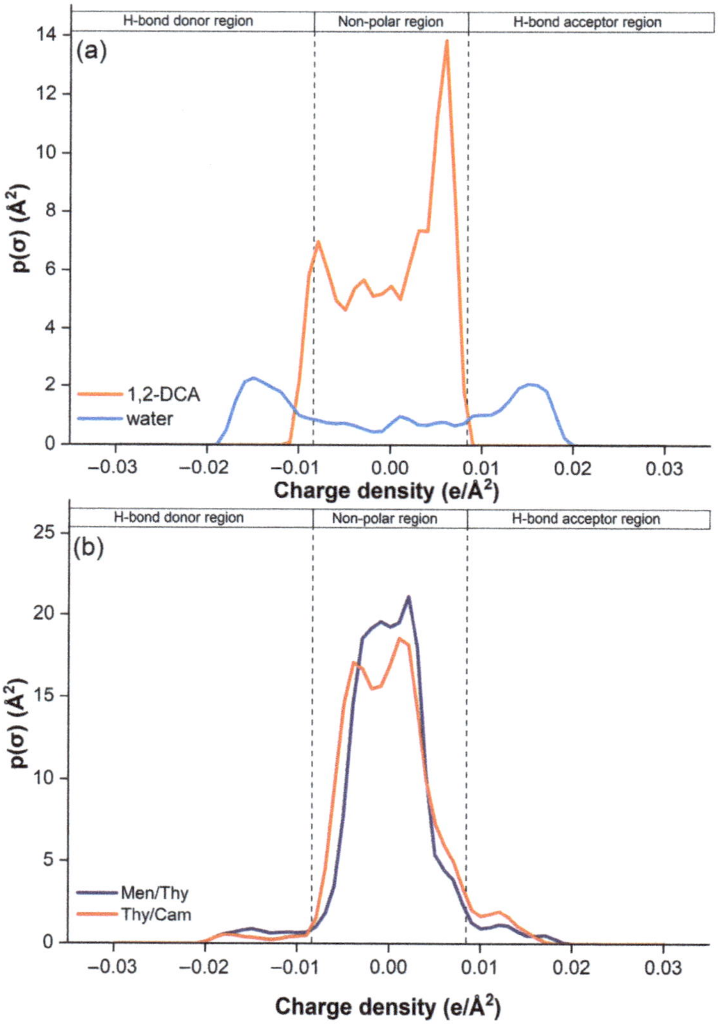

Figure 10. COSMO-RS-predicted sigma profiles of (**a**) 1,2-DCA and water, and (**b**) DESs.

Figure 10a first shows that 1,2-DCA has a relatively narrow sigma distribution dominated by non-polar and weakly polar regions, with a negligible capacity for hydrogen bonding. In contrast, water shows a very sharp and narrow peak in the negative sigma region, indicating its strong hydrogen bond donor (HBD) character. On the other hand, Figure 10b shows that both Thy/Cam and Men/Thy exhibit a broader sigma distribution extending into both the HBA and the HBD regions. This reflects their amphiphilic nature, possessing both hydrophilic (polar) and lipophilic (non-polar) properties, as well as their ability to participate in a range of non-covalent interactions, which is critical for efficient extraction of weakly polar compounds such as 1,2-DCA [49]. Indeed, Figure 11a confirms this characteristic where the sigma potential of 1,2-DCA is rather flat and close to zero, indicating a weak affinity for polar environments and weak hydrogen bonding. Water, on the other hand, has a steep sigma potential, which favors strongly polar environments and underlines its high HBD character.

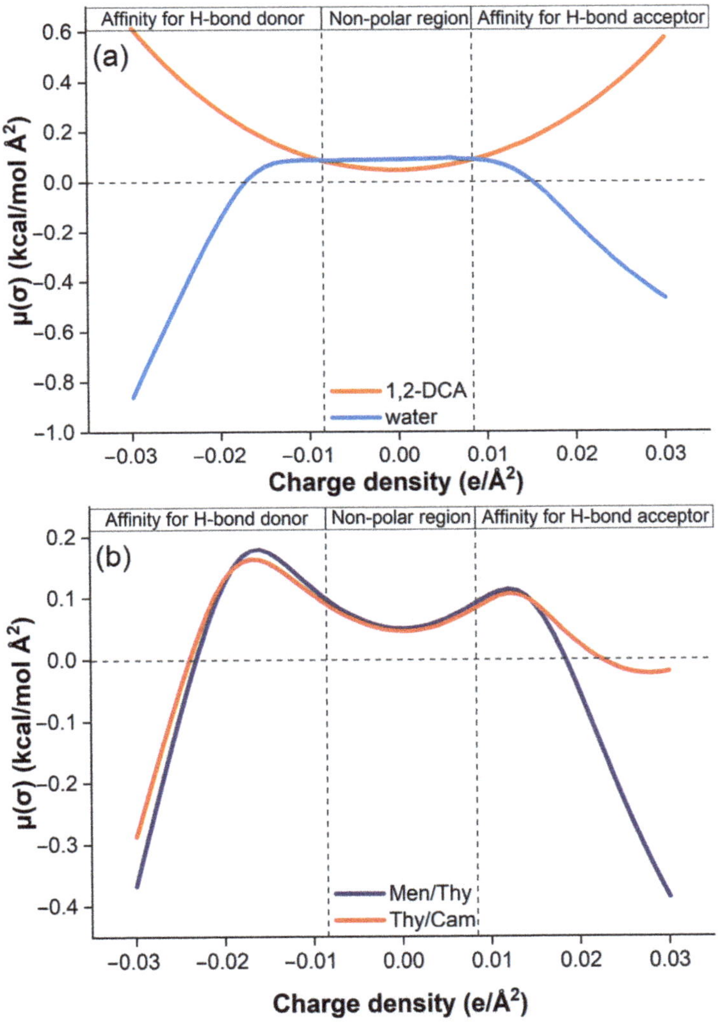

Figure 11. COSMO-RS-predicted sigma potentials of (**a**) 1,2-DCA and water, and (**b**) DESs.

Figure 11b also illustrates the interaction potential of the two HDESs. Both Thy/Cam and Men/Thy show more favorable sigma potentials over a wider range of σ values, especially in the non-polar and weakly polar regions. This confirms their compatibility with 1,2-DCA, such that strong van der Waals and dispersion interactions dominate the extraction process.

Furthermore, the good extraction performance of the terpene-based DESs obtained in this work can be explained by the molecular structures of the DES components. In fact, thymol contains an aromatic ring and a hydroxyl group, which enable both hydrophobic interactions and hydrogen bonding. Camphor, a bicyclic ketone, contributes with its strong hydrophobicity to enhance the ability of Thy/Cam DES to interact with and solubilize 1,2-DCA. Similarly, menthol, which is present in the Men/Thy DES, introduces a flexible, aliphatic hydroxyl structure that interacts with thymol to create a less viscous and more dynamic microenvironment.

Compared to other HDESs, the broad sigma profiles and moderate sigma potentials of Men/Thy and Thy/Cam enable more effective uptake of 1,2-DCA molecules, which is mainly driven by hydrophobic and dispersion forces. Their ability to solubilize 1,2-DCA while excluding water (due to minimal hydrophilic domains) leads to superior extraction efficiency, which is confirmed by the previously reported experimental results.

In summary, the qualitative results of COSMO-RS confirm the experimental performance of HDESs. The unique molecular properties and interaction profiles of Men/Thy and Thy/Cam make them highly effective for the extraction of weakly polar pollutants such as 1,2-DCA from aqueous environments, which is consistent with the environmentally friendly and sustainability goals of this study.

4. Conclusions

This study systematically evaluated the performance of solvents known as HDESs in extracting 1,2-DCA from wastewater. After screening six HDES systems, menthol/thymol (Men/Thy) and thymol/camphor (Thy/Cam) were found to be the most effective, leading to further investigation of their performance under different operating conditions. A detailed study of the parameters revealed that both systems achieved optimal extraction efficiency within 15 min, indicating rapid equilibrium and suitability for time-sensitive processes. The study also found that a 1:1 mass ratio of HDES to water was the most efficient, allowing for high extraction performance while minimizing solvent use, which is crucial for sustainable operation in practical applications.

Further analysis of the HDES molar ratios revealed that a 1:1 ratio was optimal for both Men/Thy and Thy/Cam systems. Deviations from this balance resulted in decreased efficiency due to disruptions in the solvent's molecular interactions, which are critical for the selective capture of 1,2-DCA. The study also found that higher initial concentrations of 1,2-DCA in the wastewater led to increased extraction efficiencies. The study suggests that as the contaminant concentration increases, the driving force for mass transfer strengthens, leading to improved partitioning into the HDES phase. These findings demonstrate that both Men/Thy and Thy/Cam HDES systems exhibit robust performance across a range of conditions, making them viable options for large-scale industrial applications in wastewater treatment. Unlike traditional solvents, these HDESs combine high efficiency with environmental sustainability, offering a greener alternative for the chemical industry. It is worth noting that although the regeneration of HDESs is not addressed in this work, it can be easily performed by distillation due to the large difference between the boiling points of 1,2-dichloroethane (83.5 °C) and the individual HDES components (menthol: 212 °C, thymol: 232 °C, and camphor: 209 °C). The rotary evaporator can be used on a laboratory scale.

In conclusion, although this study makes a significant contribution to environmentally friendly solvent-based contaminant extraction, the transition from laboratory scale to industrial application still requires more comprehensive studies, especially in terms of HDES recovery, toxicity, and techno-economic feasibility. Future work should focus on reconciling the development of HDESs with industrial process integration, circular economy principles, and regulatory requirements to ensure both performance and sustainability.

Author Contributions: Conceptualization, I.W. and M.K.H.-K.; methodology, I.W., A.O., and L.E.B.; software, I.W. and M.K.H.-K.; validation, I.W., E.A., and M.K.H.-K.; formal analysis, I.W. and L.E.B.; investigation, M.K.H.-K., L.E.B., and I.W.; resources, E.A. and M.K.H.-K.; data curation, I.W., E.A., and M.K.H.-K.; writing—I.W. and M.K.H.-K.; writing—review and editing, I.W. and M.K.H.-K.; supervision, M.K.H.-K.; project administration, E.A., S.M., and M.K.H.-K.; funding acquisition, E.A., S.M., and M.K.H.-K. All authors have read and agreed to the published version of the manuscript.

Funding: This research was funded by the National Plan for Science, Technology and Innovation (NPSTI, KACST) of Saudi Arabia, which supported this work through project number 14-ENV-1934-02.

Data Availability Statement: The original contributions presented in this study are included in the article. Further inquiries can be directed to the corresponding author.

Conflicts of Interest: The authors declare no conflict of interest.

References

1. Larsen, T.A.; Hoffmann, S.; Lüthi, C.; Truffer, B.; Maurer, M. Emerging solutions to the water challenges of an urbanizing world. *Science* **2016**, *352*, 928–933. [CrossRef]
2. Al-Baldawi, I.A. Removal of 1, 2-Dichloroethane from real industrial wastewater using a sub-surface batch system with *Typha angustifolia* L. *Ecotoxicol. Environ. Saf.* **2018**, *147*, 260–265. [CrossRef]
3. Jeong, W.-G.; Kim, J.-G.; Baek, K. Removal of 1, 2-dichloroethane in groundwater using Fenton oxidation. *J. Hazard. Mater.* **2022**, *428*, 128253. [CrossRef]
4. Wu, S.; Wang, P.; Fang, N.; Wang, Y.; Ding, S.; Chu, Y. Insight into catalytic performance and reaction mechanism for 1,2-dichloroethane elimination over MnCoOx-supported catalyst. *Fuel* **2024**, *360*, 130551. [CrossRef]
5. Yazdanbakhsh, A.R.; Jaafarzadeh, N.; Hematian, S.; Sheikhmohammadi, A. Removal of 1,2-dichloroethane from industrial wastewater with membrane filtration. *J. Health Field* **2014**, *2*, 8–15.
6. Motomizu, S.; Sakai, T. On-line sample pretreatment: Extraction and preconcentration. In *Comprehensive Analytical Chemistry*; Elsevier: Amsterdam, The Netherlands, 2008; Volume 54, pp. 159–201.
7. Zhou, Z.; Huang, J.; Danish, M.; Zeng, G.; Yang, R.; Gu, X.; Ali, M.; Lyu, S. Insights into enhanced removal of 1, 2-dichloroethane by amorphous boron-enhanced Fenton system: Performances and mechanisms. *J. Hazard. Mater.* **2021**, *420*, 126589. [CrossRef]
8. Lee, H.-H.; Hirano, Y.; Murayama, N.; Matsumoto, S.; Shibata, J. Adsorption properties of activated carbon prepared from waste beer lees by KOH activation and CO_2 activation. *Resour. Process.* **2007**, *54*, 19–24. [CrossRef]
9. Reyes, J.; Arriola, E.; Hernández, J.A.; Fuentes, G.A. Liquid-phase chloroform hydrodechlorination catalyzed by Pd/TiO_2–Na. *Appl. Catal. A Gen.* **2015**, *497*, 211–215.
10. Baric, M.; Majone, M.; Beccari, M.; Papini, M.P. Coupling of polyhydroxybutyrate (PHB) and zero valent iron (ZVI) for enhanced treatment of chlorinated ethanes in permeable reactive barriers (PRBs). *Chem. Eng. J.* **2012**, *195*, 22–30. [CrossRef]
11. Nunez Garcia, A.; Boparai, H.K.; O'Carroll, D.M. Enhanced dechlorination of 1,2-dichloroethane by coupled nano iron-dithionite treatment. *Environ. Sci. Technol.* **2016**, *50*, 5243–5251. [CrossRef]
12. Eydivand, S.; Nikazar, M. Degradation of 1,2-dichloroethane in simulated wastewater solution: A comprehensive study by photocatalysis using TiO_2 and ZnO nanoparticles. *Chem. Eng. Commun.* **2015**, *202*, 102–111. [CrossRef]
13. Büyüksönmez, F.h.; Hess, T.F.; Crawford, R.L.; Watts, R.J. Toxic effects of modified Fenton reactions on Xanthobacter flavus FB71. *Appl. Environ. Microbiol.* **1998**, *64*, 3759–3764. [CrossRef]
14. Bello, M.M.; Raman, A.A.A.; Asghar, A. A review on approaches for addressing the limitations of Fenton oxidation for recalcitrant wastewater treatment. *Process Saf. Environ. Prot.* **2019**, *126*, 119–140. [CrossRef]
15. Liu, X.; Vellanki, B.P.; Batchelor, B.; Abdel-Wahab, A. Degradation of 1, 2-dichloroethane with advanced reduction processes (ARPs): Effects of process variables and mechanisms. *Chem. Eng. J.* **2014**, *237*, 300–307. [CrossRef]
16. Capodaglio, A.G. Critical perspective on advanced treatment processes for water and wastewater: AOPs, ARPs, and AORPs. *Appl. Sci.* **2020**, *10*, 4549. [CrossRef]
17. De Marines, F.; Cruciata, I.; Di Bella, G.; Di Trapani, D.; Giustra, M.G.; Calabrisotto, L.S.; Lucchina, P.G.; Quatrini, P.; Viviani, G. Degradation of 1,2-dichloroethane in real polluted groundwater by using enriched bacterial consortia in aerobic and anaerobic laboratory-scale conditions. *Int. Biodeterior. Biodegrad.* **2023**, *183*, 105644. [CrossRef]

18. Wu, Z.; Yu, X.; Liu, G.; Li, W.; Lu, L.; Li, P.; Xu, X.; Jiang, J.; Wang, B.; Qiao, W. Sustained detoxification of 1, 2-dichloroethane to ethylene by a symbiotic consortium containing *Dehalococcoides* species. *Environ. Pollut.* **2023**, *325*, 121443. [CrossRef]
19. Draga, M.; Ciucanu, I.; Agotici, V.; Fernandez, A.; Barna, R. Quantitative evaluation of perchlorethylene in groundwater before and after its oxidation by helical solid sorbent extraction and gas chromatography. *Chemosphere* **2009**, *75*, 1210–1214. [CrossRef]
20. Dolatabadi, M.; Ghorbanian, A.; Ahmadzadeh, S. Degradation of high-concentration of perchloroethylene from aqueous solution using electro-fenton process. *J. Environ. Health Sustain. Dev.* **2022**, *7*, 1676–1683. [CrossRef]
21. Liu, X.; Wu, M.; Zhao, J. Removal of trichloroethylene from water by bimetallic Ni/Fe nanoparticles. *Water* **2022**, *14*, 161 [CrossRef]
22. Dominguez, C.M.; Rodriguez, V.; Montero, E.; Romero, A.; Santos, A. Methanol-enhanced degradation of carbon tetrachloride by alkaline activation of persulfate: Kinetic model. *Sci. Total Environ.* **2019**, *666*, 631–640. [CrossRef]
23. Li, B.; Li, L.; Lin, K.; Zhang, W.; Lu, S.; Luo, Q. Removal of 1, 1, 1-trichloroethane from aqueous solution by a sono-activated persulfate process. *Ultrason. Sonochem.* **2013**, *20*, 855–863. [CrossRef]
24. Almomani, F.; Rene, E.R.; Veiga, M.C.; Bhosale, R.R.; Kennes, C. Treatment of waste gas contaminated with dichloromethane using photocatalytic oxidation, biodegradation and their combinations. *J. Hazard. Mater.* **2021**, *405*, 123735. [CrossRef]
25. Wang, Z.; Wu, Y.; Cao, Z.; Li, Y.; Bai, Y.; Zhang, X.; Li, T.; Ren, B. Absorption of ethylene dichloride with imidazolium-based ionic liquids. *J. Mol. Liq.* **2023**, *376*, 121449. [CrossRef]
26. Gui, C.; Li, G.; Lei, Z.; Wei, Z.; Dong, Y. Experiment and molecular mechanism of two chlorinated volatile organic compounds in ionic liquids. *Ind. Eng. Chem. Res.* **2023**, *62*, 1160–1171. [CrossRef]
27. Song, M.; Gui, C.; Lei, Z. Experimental and Simulation Studies on the Capture of Chlorinated Volatile Organic Compounds by Ionic Liquids. *Ind. Eng. Chem. Res.* **2023**, *62*, 10184–10194. [CrossRef]
28. Yang, C.; Yu, X.; Wang, L.; Shi, M.; He, G.; Li, Q. Experimental measurement and thermodynamic modelling of liquid-liquid equilibria for the separation of 1,2-dichloroethane from cyclohexane using various extractants. *J. Mol. Liq.* **2018**, *252*, 263–270. [CrossRef]
29. Rawa-Adkonis, M.; Wolska, L.; Przyjazny, A.; Namieśnik, J. Sources of errors associated with the determination of PAH and PCB analytes in water samples. *Anal. Lett.* **2006**, *39*, 2317–2331. [CrossRef]
30. Tshepelevitsh, S.; Hernits, K.; Jenčo, J.; Hawkins, J.M.; Muteki, K.; Solich, P.; Leito, I. Systematic optimization of liquid–liquid extraction for isolation of unidentified components. *ACS Omega* **2017**, *2*, 7772–7776. [CrossRef]
31. Almohasin, J.A.; Balag, J.; Miral, V.G.; Moreno, R.V.; Tongco, L.J.; Lopez, E.C.R. Green Solvents for Liquid–Liquid Extraction: Recent Advances and Future Trends. *Eng. Proc.* **2023**, *56*, 174.
32. Rudrapal, M.; Kothawade, A.P.; Ezzat, S.M.; Egbuna, C. Bioanalysis: Methods, techniques, and applications. In *Analytical Techniques in Biosciences*; Elsevier: Amsterdam, The Netherlands, 2022; pp. 1–24.
33. Wazeer, I.; Mulyono, S.; Halilu, A.; Hizaddin, H.F.; Hashim, M.A.; Hadj-Kali, M.K. Comparative analysis of lead and cadmium extraction capacities of hydrophobic deep eutectic solvents. *J. Ind. Eng. Chem.* **2025**, *143*, 303–318. [CrossRef]
34. Wazeer, I.; Hizaddin, H.F.; Wen, N.X.; El Blidi, L.; Hashim, M.A.; Hadj-Kali, M.K. Extraction of phenol as pollutant from aqueous effluents using hydrophobic deep eutectic solvents. *Water* **2023**, *15*, 4289. [CrossRef]
35. Cichowska-Kopczyńska, I.; Nowosielski, B.; Warmińska, D. Deep eutectic solvents: Properties and applications in CO_2 separation. *Molecules* **2023**, *28*, 5293. [CrossRef]
36. Lee, J.; Jung, D.; Park, K. Hydrophobic deep eutectic solvents for the extraction of organic and inorganic analytes from aqueous environments. *TrAC Trends Anal. Chem.* **2019**, *118*, 853–868. [CrossRef]
37. Van Osch, D.J.; Zubeir, L.F.; Van Den Bruinhorst, A.; Rocha, M.A.; Kroon, M.C. Hydrophobic deep eutectic solvents as water-immiscible extractants. *Green Chem.* **2015**, *17*, 4518–4521. [CrossRef]
38. Xing, J.; Liu, X.; Dai, Y.; Zhang, Y.; Su, Z.; Chen, Z.; Gao, J.; Wang, Y.; Cui, P. Phase behavior and extraction mechanism of ethanol in alcohol ester mixture separated by deep eutectic solvents. *J. Mol. Liq.* **2022**, *368*, 120694. [CrossRef]
39. Zafarani-Moattar, M.T.; Shekaari, H.; Dizaj, A.S. Investigation of solute-solvent interactions in binary and quaternary solutions containing lithium perchlorate, propylene carbonate, and the deep eutectic solvent (choline chloride/ethylene glycol) at T=(288.15 to 318.15) K. *J. Mol. Liq.* **2020**, *319*, 114090. [CrossRef]
40. Adeyemi, A.; Bergman, J.; Branalt, J.; Sävmarker, J.; Larhed, M. Continuous flow synthesis under high-temperature/high-pressure conditions using a resistively heated flow reactor. *Org. Process Res. Dev.* **2017**, *21*, 947–955. [CrossRef]
41. Wazeer, I.; Hizaddin, H.F.; Mulyono, S.; Hashim, M.A.; Hadj-Kali, M.K. Removal of cresol contaminants from aqueous media using hydrophobic deep eutectic solvents. *J. Mol. Liq.* **2024**, *407*, 125082. [CrossRef]
42. Vlachoudi, D.; Chatzimitakos, T.; Athanasiadis, V.; Bozinou, E.; Lalas, S.I. Enhanced extraction of carotenoids from tomato industry waste using menthol/fatty acid deep eutectic solvent. *Waste* **2023**, *1*, 977–992. [CrossRef]

Fan, C.; Sebbah, T.; Liu, Y.; Cao, X. Terpenoid-capric acid based natural deep eutectic solvent: Insight into the nature of low viscosity. *Clean. Eng. Technol.* **2021**, *3*, 100116. [CrossRef]

Ribeiro, B.D.; Florindo, C.; Iff, L.C.; Coelho, M.A.; Marrucho, I.M. Menthol-based eutectic mixtures: Hydrophobic low viscosity solvents. *ACS Sustain. Chem. Eng.* **2015**, *3*, 2469–2477. [CrossRef]

Martins, M.A.; Crespo, E.A.; Pontes, P.V.; Silva, L.P.; Bülow, M.; Maximo, G.J.; Batista, E.A.; Held, C.; Pinho, S.P.; Coutinho, J.A. Tunable hydrophobic eutectic solvents based on terpenes and monocarboxylic acids. *ACS Sustain. Chem. Eng.* **2018**, *6*, 8836–8846. [CrossRef]

Almustafa, G.; Sulaiman, R.; Kumar, M.; Adeyemi, I.; Arafat, H.A.; AlNashef, I. Boron extraction from aqueous medium using novel hydrophobic deep eutectic solvents. *Chem. Eng. J.* **2020**, *395*, 125173. [CrossRef]

7. Klamt, A. Conductor-like screening model for real solvents: A new approach to the quantitative calculation of solvation phenomena. *J. Phys. Chem.* **1995**, *99*, 2224–2235. [CrossRef]

8. Klamt, A.; Eckert, F. COSMO-RS: A novel and efficient method for the a priori prediction of thermophysical data of liquids. *Fluid Phase Equilib.* **2000**, *172*, 43–72. [CrossRef]

9. Stefanovic, R.; Ludwig, M.; Webber, G.B.; Atkin, R.; Page, A.J. Nanostructure, hydrogen bonding and rheology in choline chloride deep eutectic solvents as a function of the hydrogen bond donor. *Phys. Chem. Chem. Phys.* **2017**, *19*, 3297–3306. [CrossRef]

separations

MDPI

Article

The Effect of Plasma Pretreatment on the Flotation of Lithium Aluminate and Gehlenite Using Light-Switchable Collectors

Ali Zgheib [1], Maximilian Hans Fischer [1], Stéphanie Mireille Tsanang [1], Iliass El Hraoui [1], Shukang Zhang [2], Annett Wollmann [2], Alfred P. Weber [2], Ursula E. A. Fittschen [3], Thomas Schirmer [4] and Andreas Schmidt [1,*]

[1] Institute of Organic Chemistry, Clausthal University of Technology, Leibnizstraße 6, 38678 Clausthal-Zellerfeld, Germany; ali.zgheib@tu-clausthal.de (A.Z.); iliass.el.hraoui@tu-clausthal.de (I.E.H.)

[2] Institute of Particle Technology, Clausthal University of Technology, Leibnizstraße 19, 38678 Clausthal-Zellerfeld, Germany; annett.wollmann@tu-clausthal.de (A.W.); weber@mvt.tu-clausthal.de (A.P.W.)

[3] Institute of Inorganic and Analytical Chemistry, Clausthal University of Technology, Arnold-Sommerfeld-Straße 4, 38678 Clausthal-Zellerfeld, Germany; ursula.fittschen@tu-clausthal.de

[4] Institute of Repository Research, Clausthal University of Technology, Adolph-Roemer-Str. 2A, 38678 Clausthal-Zellerfeld, Germany; thomas.schirmer@tu-clausthal.de

* Correspondence: schmidt@ioc.tu-clausthal.de

Abstract: The pyridinium phenolate punicine is a switchable molecule from *Punica granatum*. Depending on the pH, punicine exists as a cation, neutral molecule, anion, or dianion. In addition, punicine reacts to light, under the influence of which it forms radical species. We report on three punicine derivatives that possess an adamantyl, 2-methylnonyl, or heptadecyl substituent and on their performance in the flotation of lithium aluminate, an engineered artificial mineral (EnAM) for the recycling of lithium, e.g., from lithium-ion batteries. By optimizing the parameters: pH and light conditions (daylight, darkness), recovery rates of 92% of $LiAlO_2$ are achieved. In all cases, the flotation of the gangue material gehlenite ($Ca_2Al[AlSiO_7]$) is suppressed. IR, the contact angle, zeta potential measurements, TG-MS, and PXRD confirm that the punicines interact with the surface of $LiAlO_2$, which is covered by $LiAl_2(OH)_7$ after contact to water, resulting in a hydrophobization of the particle. The plasma pretreatment of the lithium aluminate has a significant influence on the flotation results and increases the recovery rates of lithium aluminate in blank tests by 58%. The oxidative plasma leads to a partial dehydratisation of the $LiAl_2(OH)_7$ and thus to a hydrophobization of the particles, while a reductive plasma causes a more hydrophilic particle surface.

Keywords: lithium recycling; critical raw material; photoswitchability; flotation; plasma; collector; lithium-ion batteries; lithium; collector; Hallimond tube

1. Introduction

Energy storage is the key to enabling the power grids of the future to run on 100% renewable energy, gradually replacing fossil fuels. In this regard, lithium-ion batteries are currently one of the most important storage devices for electrical energy [1–3]. Having long dominated as power sources for consumer electronics, lithium-ion batteries are now also used in transportation, particularly in electric vehicles [4,5]. Apart from that, however, lithium is also an important strategic raw material for many other sectors and industries [6]. The world's lithium reserves, which are unequally distributed geographically, are mainly found in Chile (9,300,000 t), Australia (6,200,000 t), Argentina (3,600,000 t), and China (3,000,000 t) [7]. Their extraction is energy- and labor-intensive, resulting in significant environmental pollution [8,9]. These problems are becoming more serious due to the ever-increasing demand for lithium [10], making the development of new lithium recycling strategies highly relevant. Intensive research is currently being conducted into how lithium can be enriched in various phases to form so-called engineered artificial minerals (EnAMs) when recycling lithium-ion batteries. For example, lithium can be separated as lithium aluminate ($LiAlO_2$) from Umicore slag, based on the Li_2O-MgO-Al_2O_3-SiO_2-CaO slag system with a high aluminum content [11–13]. The crude lithium aluminate produced in this way contains up to 10.6% lithium. However, many other minerals can be formed during the pyrometallurgical recycling processes of lithium-ion batteries, and these can compete with the extraction of lithium-containing minerals; these minerals are referred to as gangue materials. Besides galaxite ($MnAl_2O_4$), which forms in the presence of manganese, gehlenite ($Ca_2Al[AlSiO_7]$) is one of the most difficult minerals to separate [14–17]. Of all the separation processes, flotation [18] is one of the most important, and new challenges in the recycling of critical raw materials also require fundamental research in this separation method. Flotation is based on differences in the wettability, density, and hydrophobicity of the desired minerals and their environment. Injected air forms small bubbles in the mineral suspension, and the hydrophobicity of the desired mineral particles is increased as selectively as possible by adding collectors. These hydrophobic particles can then attach themselves to the gas bubbles and rise to form a foam layer that overflows as a mineral-rich concentrate [18–23]. Improving the interaction between the collector and the dispersed particles is therefore one of the main optimization goals. Sodium oleate is a classic collector in this regard and has also been used for the flotation of lithium aluminate [24,25]. However, switchable molecules have an advantage over established collectors such as the one mentioned, in that their molecular and intermolecular properties can be reversibly adapted to the separation problem by simple external parameters. Therefore, the natural product punicine has recently become the focus of interest [26,27]. Punicine was isolated from the leaves of *Punica granatum* (pomegranate) and described as a cation $\mathbf{P^+}$ **[28]**. Due to its ability to stabilize the negative and positive charge within the π-electron system at a neutral pH (Scheme 1), it belongs to the substance class of heterocyclic mesomeric betaines [29]. In a neutral solution, punicine exists as a mixture of two different tautomers, $\mathbf{P^{+/-}A}$ and $\mathbf{P^{+/-}B}$. In a basic medium, the anionic tripole $\mathbf{P^-}$ is formed, in which two negative charges and one positive charge are delocalized within their common π-electron system [30]. In addition, punicines can undergo ring-opening reactions, and some derivatives can be oxidized [31]. Apart from the different properties that arise from changes in pH, punicines also react to light. They have different properties under UV irradiation than in the dark, since radicals are formed under the influence of light, which, in the simplest case, are formed by a disproportionation of two punicine betaines in the ground state and can be formulated as radical cations $\mathbf{P^{\cdot+}}$ and radical anions $\mathbf{P^{\cdot-}}$ [32].

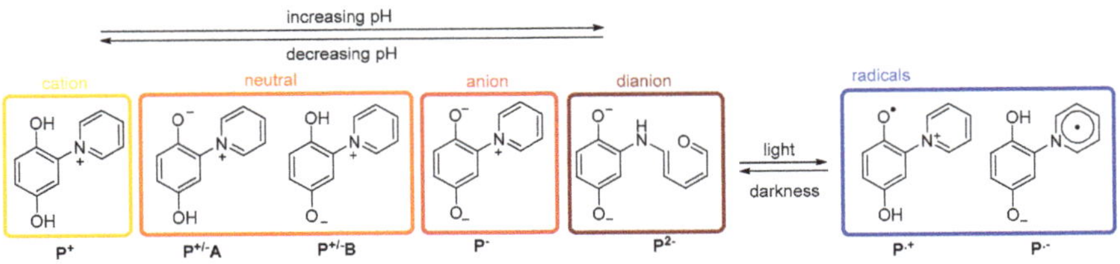

Scheme 1. Schematic representation of the switchability of punicine with respect to charges and radical status.

The punicine molecule can therefore exist in several states, into which it can be reversibly converted. As far as the intermolecular interactions are concerned, vertical interactions with itself [29] in an aqueous solution or in the crystal and with electron-rich aromatic compounds such as hydroquinone were found, which, in the latter case, led to a crystalline sandwich complex [33] (Scheme 2). Hydrogen bonds, i.e., horizontal interactions, were detected in DMSO [29]. The 2-olate group is the site that attacks electrophilic substrates such as acetylenes [34] and boron compounds [35], respectively; it also binds lithium cations [27] and minerals in the solid state [36].

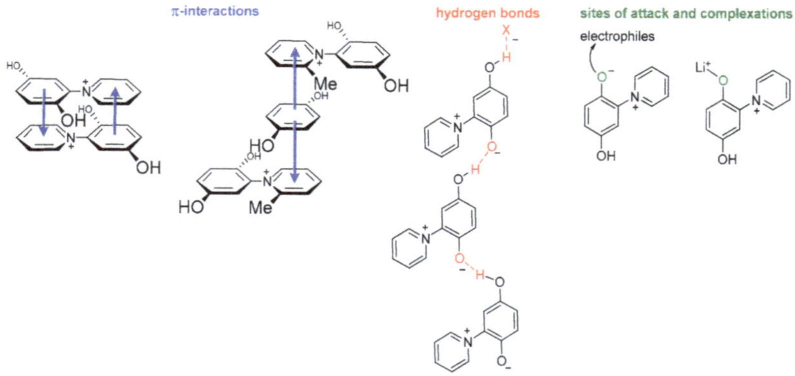

Scheme 2. Intermolecular interactions and complexes of punicine.

In this study, the froth flotation of lithium aluminate ($LiAlO_2$) and the gangue material gehlenite ($Ca_2Al[AlSiO_7]$) was carried out using punicine derivatives as collectors. On the one hand, we examined the influence of various hydrophobic groups on the flotation results. On the other hand, we examined the pretreatment of these minerals by oxidative as well as reductive plasma, and studied the effect on the recovery rates of the flotations. The flotation was carried out under a range of different light scenarios and in different pH media to take advantage of the switchability of punicines.

2. Materials and Methods

2.1. General

The light tests were carried out in a closed box, painted black on the inside, in which the Hallimond tube was installed. The box measures 60 × 60 × 60 cm and can be opened on two sides with hinges (cf. Supporting Information). In the middle of each of the four walls is an aluminum heat sink from Fischer (thermal resistance: 0.7–1.65 K/W and dimensions: 200 × 200 × 25 mm), on which four LEDs from Avonec (Premium 3W LED, 390–400 nm, color: UV-A, max. power: 750 mA, and operating voltage: 3.5 V–4.5 V) were mounted, which can be operated via a power supply unit (APC-16–700, input: 100–240 V, output: +24 V, max. 700 mA, and class 2 power supply unit). The light sources were switched on for all experiments at the beginning of the conditioning. The daylight flotations were carried out at a light intensity of >3000 lux. The UV irradiations were carried out in UV-A light with a light intensity of 4500 lux, while the light intensity in the dark was <40 lux. The required pH value was achieved by adding 0.1 M NaOH and adjusted with the Britton–Robinson buffer: pH 2.0 ± 0.3 was adjusted by adding 20 mL Britton–Robinson buffer and pH 7.0 ± 0.3 was achieved by adding 40 mL Britton–Robinson buffer and 20 mL 0.1 NaOH. The pH value 13.0 ± 0.3 was adjusted by adding 20 mL of 0.1 NaOH. In all cases, the remaining volume was made up to 250 mL.

2.2. Zeta Potential Measurements

The measurements were conducted with a Zetasizer Nano from Malvern (Kassel, Germany). The respective collector was added to the solution. An amount of 1 g of the mineral was suspended in water and allowed to settle, so that large particles sedimented. The measurements were carried out with the supernatant dispersion.

2.3. Particle Size Measurements

The particle size distribution of mixed samples and single mineral samples of $LiAlO_2$ and gehlenite was determined with dry measurement via the HELOS laser diffractometer from Sympatec (Clausthal-Zellerfeld, Germany) at an air pressure of 4 bar.

2.4. Contact Angles

The contact angles of conditioned and untreated $LiAlO_2$ and gehlenite materials were conducted with a tensiometer OCA 15plus, using the sessile drop technique (dataphysics, Filderstadt, Germany). Therefore, powder of the sample material with a particle size smaller than 100 µm was fixed on a tab by pressing with a spatula. For the measurement, a 5.000 µL water bubble with a rate of 2 µL/s was injected with the 500 µL hamilton syringe and deposited on the sample surface. For each sample, six measurements were done, using different surface positions, and the average value was determined.

2.5. Stock Solutions of the Punicines for Flotations

Stock solutions of the punicines were prepared, so that 60 µL of the stock solution in water/ethanol = 1:1 (punicines **1** and **2**) or isopropanol (punicine **3**) corresponded to a concentration of 1×10^{-5} mol/L, and 120 µL to 2×10^{-5} mol/L of punicine in a 250 mL microflotation experiment.

2.6. Syntheses

Syntheses are described in the Supporting Information, including those of starting materials [37].

3. Results

3.1. Syntheses of the Punicine Derivatives

Hydrophobic substituents of different architectures were selected for our studies. We compared a punicine with an adamantyl substituent as a rigid and bulky residue with those that are long-chain substituted. For this purpose, we chose punicines with an n-heptadecane and a branched decyl residue. The adamantyl group ($C_{10}H_{15}$) has an estimated hydrophobic substituent constant π of 3.1 [38] and the log p value, a measure of the hydrophilicity of a compound of adamantane (4.22), is orders of magnitude lower than the log p value of n-heptadecane (8.69). Adamantyl-substituted compounds are, therefore, less hydrophobic than analogous linear-substituted compounds. The choice of a branched long-chain substituent followed the example of nature, where, for example, branched fatty acids serve as highly hydrophobic biomaterials [39]. These punicines were prepared easily by mixing p-benzoquinone with the corresponding pyridine derivatives in the presence of glacial acetic acid (Scheme 3). The reaction of p-benzoquinone with 4-(adamantanyl-1-methyl)-pyridine, 4-(2-methylnonyl)-pyridine, and 4-hexadecylpyridine gave the desired punicines **1**, **2**, and **3**, respectively. The starting materials 4-adamantyl-, 4-(nonan-2-yl)- and 4-heptadecylpyridine were synthesized according to the literature [40]. Synthesic procedures are described in the Supporting Information.

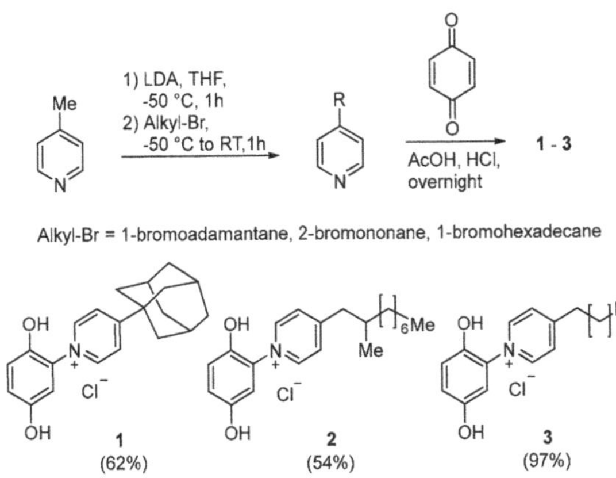

Scheme 3. Synthesis of the punicine derivatives.

3.2. Characterization of the Minerals

Before the minerals were subjected to the plasma treatment and various flotation tests, powder X-ray diffraction measurements were carried out to gain information about the polymorphs of the lithium aluminate we used. The X-ray diffractograms are shown in the Supporting Information. This is because several different phases of lithium aluminate have been found to date, which differ in their properties [41–43]. For example, lithium aluminate in the γ-phase has a tetragonal crystal structure of space group P41212 (space group no. 92), while its α-modification has a trigonal crystal structure of space group R3m (space group no. 166). Thus, under suitable conditions, only the α-phase can react with an almost complete exchange of Li^+ for H^+ [41], forming $Li_{1\text{-}x}H_x(AlO_2)$ ($x \geq 0.95$), whereas γ-$LiAlO_2$ does not [44]. This is attributed to different Li^+ electrophilicities, a significantly smaller Li-binding energy, and different Li-dislocating energies in both phases. Our measurements showed that the peaks can be assigned to γ-$LiAlO_2$. These results are in good agreement with earlier investigations of the crystal phases of these minerals [45]. As gehlenite is not produced industrially and does not occur in its pure form in nature, a rock with a high gehlenite content form Vata de Sus, Romania, had to be used. Our XRD and ICP-OES measurements revealed that it has a purity of approximately 56% and is composed of 20–24% garnet, including iron-rich grossular garnet, 17% vesuvianite, and impurities of 7% calcite and 1% quartz. This sample therefore mimics the situation in authentic slag during recycling, in terms of the variety of compositions. Since the particle sizes have a considerable influence on the flotation results, these were determined second. The x50 value for the particle size of the $LiAlO_2$ particles was 20 μm, and 90% of the particles were smaller than 60 μm. The x50 value for the particle size of gehlenite was 27 μm, and 90% of the particles were smaller than 60 μm. The particle size distribution was between 1 μm and 80 μm (for details, cf. Supporting Information). The much-cited insolubility [46] of lithium aluminate in water, a parameter that is very important for flotation, is apparently not entirely accurate and must be considered when developing new collectors. For example, an aqueous suspension of lithium aluminate has a pH of 10.9 ± 0.3, which indicates the formation of hydroxide anions in accordance with Scheme 4 [47]. The subsequent reaction of the resulting lithium aluminate hydroxide hydrate with hydroxide releases lithium cations into the aqueous phase (Scheme 4). The subsequent reactions of lithium aluminate hydroxide hydrate with CO_2 to form carbonate species such as $Li_2Al_4(CO_3)(OH)_{12} \times 3\ H_2O$, $Li_2[Al_2(OH)_6]^{2+} \times CO_3{}^{2-} \times n\ H_2O$, or Li_2CO_3 are also possible [48–52].

$$2\ LiAlO_2 + (4{+}n)\ H_2O \longrightarrow LiAl_2(OH)_7 \cdot n\ H_2O + Li^+ + OH^-$$

$$LiAl_2(OH)_7 \cdot 2\ H_2O + OH^- \longrightarrow Li^+ + 2\ Al(OH)_4^- + 2\ H_2O$$

Scheme 4. Reactions of lithium aluminate surfaces with water.

3.3. Plasma Pretreatment of the Minerals Prior to Flotation

To further investigate the effects of the surface properties of the minerals on the flotation process, $LiAlO_2$ and the gehlenite-rich material were separately subjected to plasma pretreatment. The experiments with oxidative plasma were carried out at different time intervals in a small borosilicate glass tube with a diameter of 2.5 cm. The tube was connected to the gas cylinder filled with air at one end, the other end was left open to allow the gas to flow at an air flow rate of 2 mL/min, and 2 g of the desired mineral was added to the center (for details, see Supporting Information). Under the influence of a high voltage (160 V) and a frequency of 106.4 kHz, a strong electromagnetic field is generated, which in turn ionizes the gas flow (air) and creates the plasma state. The plasma then comes into contact with the mineral surface for the selected different time intervals [53]. The temperature inside the tube containing the sample was measured during and after the plasma treatment and found to be around 30 °C, so that the observed effects of the plasma are not consequences of the thermal treatment. It is known that such a generated oxidative plasma causes the formation of oxide layers on some mineral surfaces, such as chalcopyrite and molybdenite [54]. Figure 1 shows the thermal stability of 5 min plasma-pretreated $LiAlO_2$ and a sample without any plasma pretreatment. Different phase transition temperatures are observed in both samples. The untreated sample shows a significant decrease in mass in the temperature range from 150 °C to 350 °C. The curve flattens out at higher temperatures. The sample that was pretreated with plasma for five minutes shows a much smaller decrease in sample mass overall; the decrease in the temperature ranges from 150 to 350 °C is less than 0.5%. In comparison to the literature [48], and in view of the beforementioned reaction of $LiAlO_2$ with water, the observed decrease in mass can be attributed to the loss of water from species such as $LiAl_2(OH)_7 \times n\ H_2O$. The consequence is that a hydrophobic surface is at least partially re-formed by the plasma treatment, which is indeed reflected in the thermogravimetric analysis, IR investigations, and the flotation results (see below).

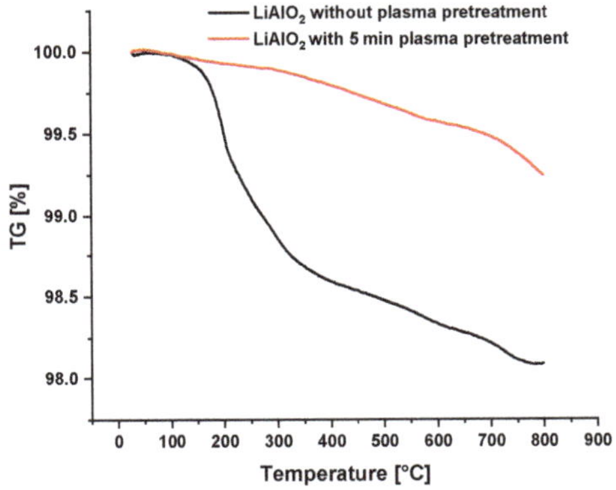

Figure 1. TGA measurements of pure lithium aluminate and 5 min oxidative plasma-pretreated lithium aluminate samples.

The results of the thermogravimetric analysis coupled with mass spectrometry (TG-MS) are shown in Figure 2. The results confirm that treatment with oxidative plasma causes a partial dehydratization of the $LiAl_2(OH)_7$ on the surface of the $LiAlO_2$ particles. In the $LiAlO_2$ sample, water molecules are released when heated to 300 °C, in addition to the release of a small amount of CO_2 molecules at 750 °C ($<5.89 \times 10^{-11}$ A), which originate from the carbonates such as $Li_2Al_4(CO_3)(OH)_{12} \times 3\ H_2O$, $Li_2[Al_2(OH)_6]^{2+} \times CO_3^{2-}$, or Li_2CO_3 on the surface mentioned in Section 3.1 [48–52]. In the plasma-pretreated sample, which was exposed to the plasma for 5 min, fewer OH groups were found due to the beforementioned dehydratization, and instead, the decarboxylations become more prominent (8.21×10^{-11} A) [55]. In addition, NO groups were also found on the surface of the plasma-pretreated samples, originating from the air used for the plasma generation [56].

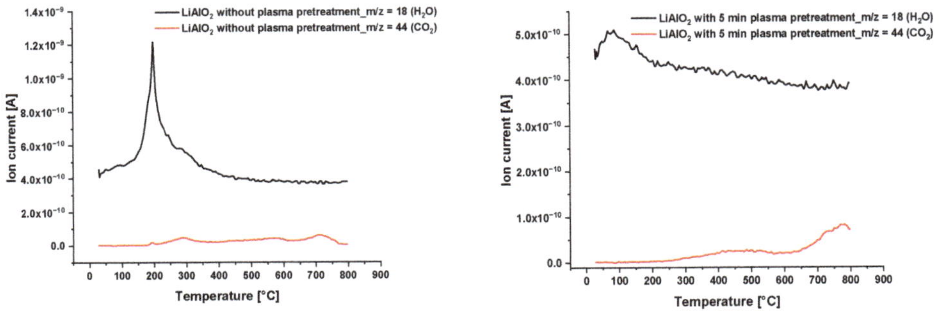

Figure 2. TG-MS of pure $LiAlO_2$ (**left**) and 5 min plasma pretreated $LiAlO_2$ (**right**).

3.4. Flotations

3.4.1. Flotation of Lithium Aluminate Without Plasma Pretreatment

In order to determine the influence of the plasma pretreatment on the flotations, blank experiments were initially carried out. Thus, lithium aluminate was floated by microflotation tests using a Hallimond tube, the design of which is outlined in the Supporting Information. The flotation tests were carried out in the presence of the punicines **1**, **2**, and **3** as collectors, respectively, and pine oil as a foaming agent. Punicine **1** and **2** were dissolved in a 1:1 fraction of water/ethanol, while punicine **3** was dissolved in isopropanol. The effectiveness of the collectors was tested by varying the flotation parameters, mainly by changing the pH of the medium and the concentration of the collectors (cf. Figure 3). In addition, microflotations were carried out under different lighting scenarios, with flotation taking place in daylight (>3000 lux), UV-light (λ = 390–400 nm, 4500 lux), and darkness (<40 lux). All measurements were repeated five to six times. We determined the average values and the confidence intervals. In the experiments, samples of 2.00 g of the desired mineral were subjected to flotation, and the yield was calculated from the mass ratio of concentrate and residue. The specifications of the light sources, including the emission spectra, can be found in the Supporting Information.

To study suitable flotation conditions with punicines **1**–**3**, blank tests were first carried out. Thus, $LiAlO_2$ without a collector was floated under different light conditions to see if light has any influence on the mineral itself. Our results show that this is not the case. The flotation of $LiAlO_2$ in daylight gives a yield of 26.53% ± 3.00, with identical values obtained under UV irradiation and in the dark, within the error limits. As can be seen from Figure 3, the punicines cause significant light dependencies. For example, while the use of punicine **1** in daylight hardly improves the flotation results, there is an increase to 42.3% (i.e., by 53%) in the dark. Replacing punicine **1** with punicine **2** even leads to an increase of 153% to 69.98 ± 3.27% under the same conditions. Overall, punicine **3** shows the lowest light dependence, but leads to the highest yields with 88.15% ± 7.61%. The results show that the hydrophilicity index log P, as mentioned above, is indeed related to the flotation results, as the value of the adamantyl residue indicates less hydrophobic properties. A comparison with data from n-decyl-substituted punicine [36], which is the isomer of punicine **2**, shows that the effect of α-branching of the side chain has no significant effect on the recovery rate.

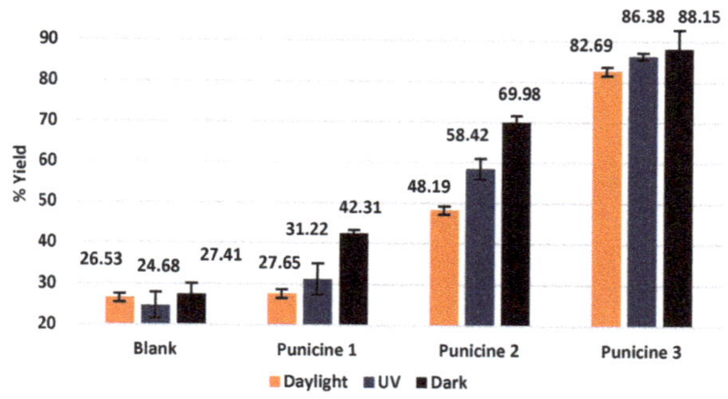

Figure 3. Comparison of the recovery rates of lithium aluminate at different light scenarios in the presence of the collectors **1**–**3** under identical conditions, i.e., 2.00 g of $LiAlO_2$, 1 min of mixing (500 rpm) with 25 mL of distilled water, 60 μL (1×10^{-5} M) of the corresponding collector, 1 min of mixing, 30 μL of frother, and another 1 min of mixing. The flotation was done for 3 min at an air flow rate of 32 mL/min using 250 mL distilled water in daylight at a pH of 10.9 ± 0.3.

As mentioned before, the properties of punicines change not only in the light, but also by varying the pH value. Figure 4 summarizes the results of flotations at pH values 2.0 ± 0.3, 7.0 ± 0.3, and 13.0 ± 0.3, in daylight. Punicine **1** gives the best yields in acidic conditions and the worst in neutral conditions. In contrast, the yield of punicines **2** and **3** increases with the pH value and reaches a maximum of 84.14 ± 4.97% for punicine **3** at a pH of 13.0 ± 0.3, i.e., when the punicine exists as an anionic molecule. The recovery yield is, therefore, increased by 96% if the pH value is changed from 2.0 ± 0.3 to 13.0 ± 0.3. This effect is i.a. due to the increased nucleophilicity of the punicine's oxygen atoms, which are deprotonated in the alkaline. These compete with water in the reaction with the surface of the lithium aluminate.

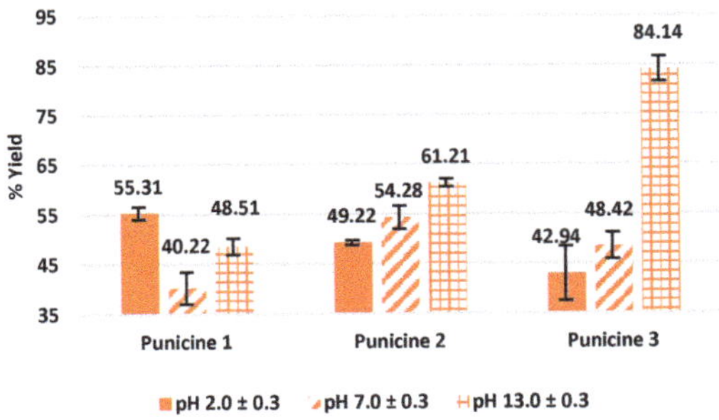

Figure 4. Comparison of the recovery rates of lithium aluminate at different pH values in the presence of the collectors **1–3** under identical conditions, i.e., 2.00 g of $LiAlO_2$, 1 min of mixing (500 rpm) with 25 mL of distilled water, 60 µL (1×10^{-5} M) of the corresponding collector, 1 min of mixing, 30 µL of frother, and another 1 min of mixing. The flotation was done for 3 min at an air flow rate of 32 mL/min using 250 mL distilled water in daylight.

3.4.2. Flotation of the Gangue Material Gehlenite in Comparison to Lithium Aluminate Without Plasma Pretreatment

The flotation of the gangue mineral gehlenite was also carried out and compared with the lithium-bearing minerals under the same conditions. Figure 5 shows the flotation results of gehlenite compared to $LiAlO_2$ using 2×10^{-5} mol/L punicine **3**. The percentage recovery of gehlenite under different light conditions was approximately the same, ranging from 20.50 ± 2.28% in daylight to 21.76 ± 1.67% under UV irradiation and 19.57 ± 4.08% in the dark. The same flotation conditions were carried out with $LiAlO_2$. Due to the increased collector concentration compared to the results shown before, a very high recovery rate of $LiAlO_2$ was achieved (90.35 ± 3.22%). The same experiments were performed with punicine **2** as a control, showing a very similar behavior. Approximately the same percentage recovery of gehlenite was obtained (21.83 ± 5.41% in daylight), while the percentage recovery of $LiAlO_2$ in daylight was 70.37 ± 2.47% (see Supporting Information).

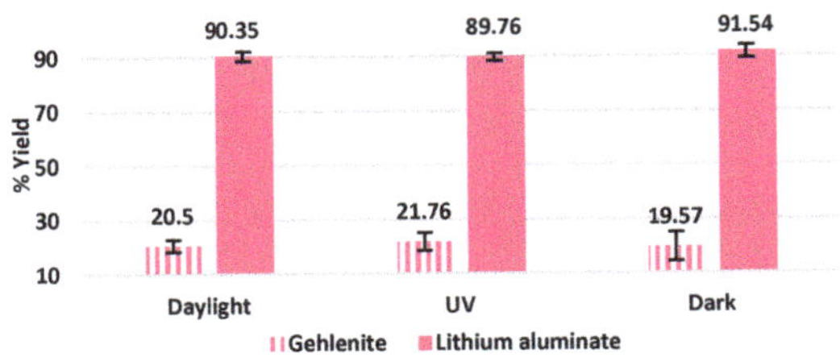

Figure 5. Single mineral flotation of lithium aluminate and gehlenite using punicine **3** as collector at pH 10.9 ± 0.3 for lithium aluminate and pH 8.3 ± 0.3 for gehlenite at different light scenarios. The conditioning was as follows: 120 µL collector, 1 min mixing, 30 µL frother, 1 min mixing, stirring speed 500 rpm, air flow rate of 32 cm^3/min using 250 mL distilled water, and flotation time of 3 min.

3.4.3. Flotations After Oxidative and Reductive Plasma Pretreatment

Figure 6 shows the results of the blank flotation of $LiAlO_2$ and gehlenite as well as of samples, which were pretreated with plasma for different time intervals (1, 3, and 5 min). The flotation experiments were carried out immediately after the plasma pretreatment, and each experiment was repeated at least three times. Only very slight changes in the flotation of plasma-pretreated gehlenite-rich material were observed compared to that without the plasma pretreatment. However, for $LiAlO_2$, the percent yield increased from 26.52 ± 1.50% to 42.08 ± 1.18% when the latter was treated with plasma for 1 min. This is consistent with the previously mentioned results of the thermogravimetric analyses and can be explained by the partial dehydration of $LiAl_2(OH)_7 \times n\ H_2O$ and a partial re-formation of a more hydrophobic surface. Beyond that, a longer plasma treatment does not increase the yield of the flotation.

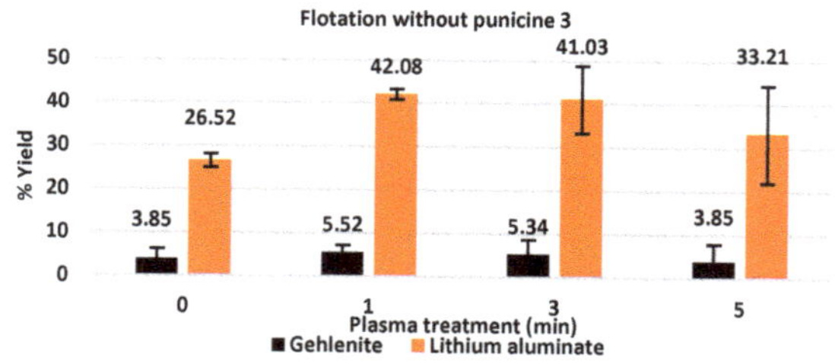

Figure 6. Single mineral flotation of plasma-pretreated lithium aluminate and the gehlenite-rich material without collector at pH 10.9 ± 0.3 for lithium aluminate and pH 8.3 ± 0.3 for gehlenite as a function of plasma treatment time. Conditioning was as follows: 30 μL frother, 1 min mixing, stirring speed 500 rpm, air flow rate of 32 cm^3/min using 250 mL distilled water, and flotation time of 3 min.

The flotation of the plasma-treated minerals was then carried out with punicine **3** as the collector (Figure 7), which had been found to be the best in the previous tests. Again, no significant effect on the percent recovery of gehlenite was observed, regardless of the duration of the plasma pretreatment. For $LiAlO_2$, however, the percent recovery gradually increased with the duration of the plasma pretreatment, reaching a maximum of 81.29 ± 3.26% yield after 3 min. The presence of punicine **3** thus almost doubles the yield compared to the blank samples.

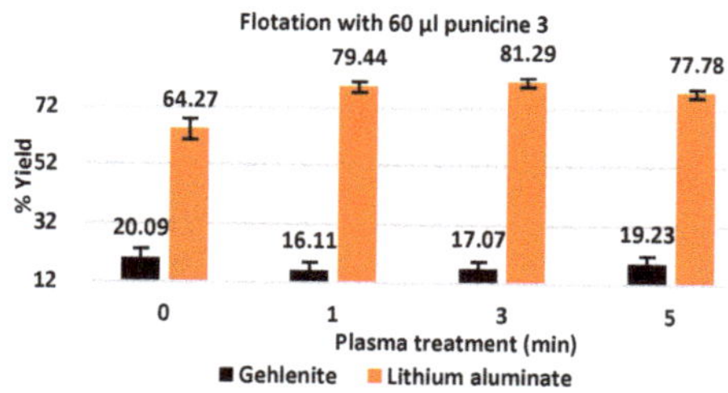

Figure 7. Single mineral flotation of the oxidative plasma pretreated lithium aluminate and gehlenite using punicine **3** as a collector at pH 10.9 ± 0.3 for lithium aluminate, and pH 8.3 for gehlenite as a function of plasma treatment time in min. The conditioning was as follows: 60 μL collector, 1 min mixing 30 μL frother, 1 min mixing, stirring speed of 500 rpm, air flow rate of 32 cm^3/min using 250 mL distilled water, and applying a flotation time of 3 min.

As a comparison, the influence of reductive plasma on the flotation results was also investigated. For this purpose, $LiAlO_2$ samples were pretreated with plasma for 1 min, 3 min, and 5 min, respectively, using Ar/H_2 as the gas for the plasma experiment. In this case, the plasma was generated at 90 V. The samples were then floated with punicine **3** as a collector and compared with blank samples. The percentage recoveries are shown in Figure 8. The percentage recovery rates obtained by the flotation of Ar/H_2-plasma-treated $LiAlO_2$ indicate that the reducing plasma has an overall weaker effect on the flotation process than the oxidizing plasma. In the blank flotation without the collector punicine **3**, the percent recovery of the pretreated samples initially increased from 16.22 ± 6.16% to 19.11 ± 14.27%. The percentage recoveries decreased to their lowest value (13.23 ± 2.11) when the plasma pretreatment time was further increased to 5 min. When punicine **3** was used as a collector during flotation, a similar increasing profile was observed. Increasing the plasma pretreatment time from 0 to 3 min increased the percentage recovery from 36.54 ± 11.48% to 43.87 ± 7.52%. The percentage recovery decreased to 34.62 ± 4.39% when the plasma pretreatment time was increased to 5 min.

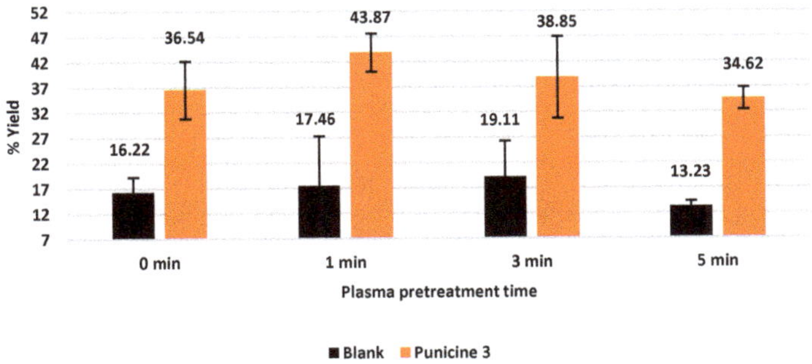

Figure 8. Single mineral flotation of the reductive plasma pretreated lithium aluminate in blank and using punicine **3** as a collector at pH 10.9 ± 0.3 as a function of plasma treatment time in min. The conditioning was as follows: 60 μL collector, 1 min mixing 30 μL frother, 1 min mixing, stirring speed of 500 rpm, air flow rate of 32 cm^3/min using 250 mL distilled water, and flotation time of 3 min.

4. Discussion

The flotation results show that punicines lead to a significant hydrophobization of the lithium aluminate particle surface. This is also reflected in the contact angle measurements that we carried out for lithium aluminate and gehlenite before and after flotation with punicines **1**, **2**, and **3** (Figure 9). The results show a low contact angle for pure $LiAlO_2$ (16.06° ± 2.06°), which confirms the hydrophilic surface of the mineral, which, as mentioned at the beginning, is i.a. due to the formation of lithium aluminum hydroxide hydrate with water on the surface of the particles (cf. Figure 6). In the case of the gehlenite-rich material used here, the measurements show a higher contact angle (86.03° ± 4.54°), which indicates a hydrophobic surface a priori, and this result is in good agreement with the literature [40]. In the case of $LiAlO_2$, the contact angle after flotation with 120 μL of the stock solution of punicine **3** is 103.84° ± 10.87° due to a considerable hydrophobization of the particle surface, while in the case of gehlenite, only a slight increase in the contact angle was observed (106.66° ± 1.45°). Similar experiments were carried out in the presence of punicines **1** and **2**. The use of punicine **1** slightly increases the contact angle of $LiAlO_2$ to 31.93° ± 5.92°. A similar result is obtained with punicine **2**, where the contact angle increases to 36.97° ± 0.68° after flotation. The sequence of contact angles is therefore roughly parallel to the flotation results shown in Figure 3 (vide supra).

The hydrophobization of the lithium aluminate surface can be attributed, at least in part, to interactions of punicines with the lithium aluminum hydroxide hydrate $LiAl_2(OH)_7$, which is known, on the one hand, to dissociate to form hydroxide and $LiAl_2(OH)_6^+$ cations, and on the other hand, to replace complexed hydroxides toward a variety of nucleophiles [51]. As a result, the interaction with punicines can be understood as a complex formation according to (Scheme 5) and a Coulomb interaction by the formation of a salt as expressed by (Scheme 5). The punicines bound in this way have an OH group that is acidic and can be completely deprotonated in an alkaline environment, so that the particle surface is covered either with cationic collector molecules or with neutral ones. This explains the strong pH dependence of the flotation results shown in Figure 4. Since our PXRD measurements confirmed that the $LiAlO_2$ used is present in the γ-phase, Li^+/H^+ exchange reactions in acid under formation of $Li_{1-x}H_x(AlO_2)$ [45] can be excluded as the reason for the different flotation behavior. The photoswitchability also remains unchanged, as the results in Figure 3 show.

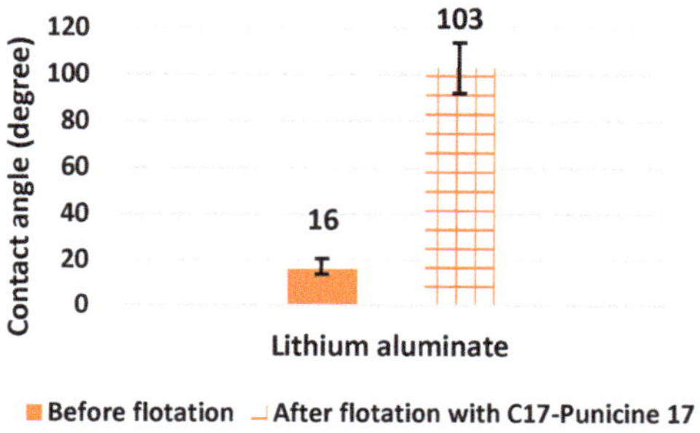

Figure 9. Contact angle on $LiAlO_2$ before and after flotation with punicine **3** under identical conditions, i.e., 2.00 g of the mineral, 1 min mixing (500 rpm) with 25 mL distilled water, 120 μL (2×10^{-5} M) of the corresponding collector, 1 min mixing, 30 μL foaming agent, and 1 min mixing again. Flotation was carried out for 3 min at an air flow rate of 32 mL/min with 250 mL of distilled water in daylight and a pH of 10.9 ± 0.3.

The electrical potential at the sliding surface of the particle surface of the minerals in suspension was measured (Figure 10). Two series of measurements were performed, the first considering the mineral $LiAlO_2$, and the second with the addition of 0.06 mM of punicines **1**, **2**, and **3** to the mineral, respectively, in a pH range of 2.0 ± 0.3, 4.0 ± 0.3, 7.0 ± 0.3, 8.0 ± 0.3, 11.0 ± 0.3, and 13.0 ± 0.3. Figure 10 shows the zeta potentials as functions of the pH. The curve of pure lithium aluminate cross at zero zeta potential at about a pH of 4.2, which is the point of zero charge (PZC) of the material used in this study. Under extremely acidic conditions, $LiAlO_2$ shows a positive zeta potential (+32.16 mV), which increases to +39.50 mV, +45.06 mV, and +40.87 mV when punicine **1**, **2**, and **3** are added, respectively. At this pH, the punicines are also cationic, so the increase in zeta potential indicates a chemosorped localization of the cationic surfactants at the mineral surface. By increasing the pH of the medium, the zeta potential of $LiAlO_2$ gradually decreased. At a pH of 4.0 ± 0.3, the zeta potential of pure $LiAlO_2$ was +3.56 mV, which decreased to −30.46 mV when punicine **1** was added and to −29.33 mV when punicine **2** was added. When punicine **3** was added, a smaller decrease in the zeta potential of $LiAlO_2$ at a pH of 4.0 ± 0.3 was observed (−1.84 mV). The zeta potential of $LiAlO_2$ shows a minimum value at a pH of 8.0 ± 0.3 (−33.80 mV). The addition of punicine **1** and **2** resulted in a decrease in the zeta potential (−62.16 mV and −56.36 mV, respectively).

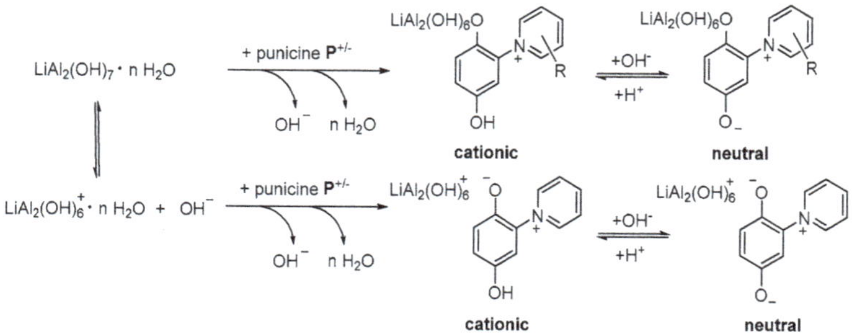

Scheme 5. Interactions of punicines with lithium aluminum surfaces.

When punicine **3** was added, a huge shift in the zeta potential was observed, where a positive zeta potential was measured (+1.82 mV). This measurement was repeated three times, and similar positive values were obtained in all three cases. This result is consistent with results from the literature [27], where an increase in zeta potential was noted at a pH of 8.0 ± 0.3 when long-chain punicines were used. At a higher pH, a slight increase in zeta potential was observed. The zeta potential of $LiAlO_2$ at different pH conditions was also determined using different collector concentrations (Supporting Information), using

0.02 mM and 0.04 mM of punicine **3**. A slight decrease in the zeta potential of $LiAlO_2$ was detected with increasing concentrations of punicine **3**. The same effect on the zeta potential was found at pH 7.0 ± 0.3 with lower concentrations of punicine **3**.

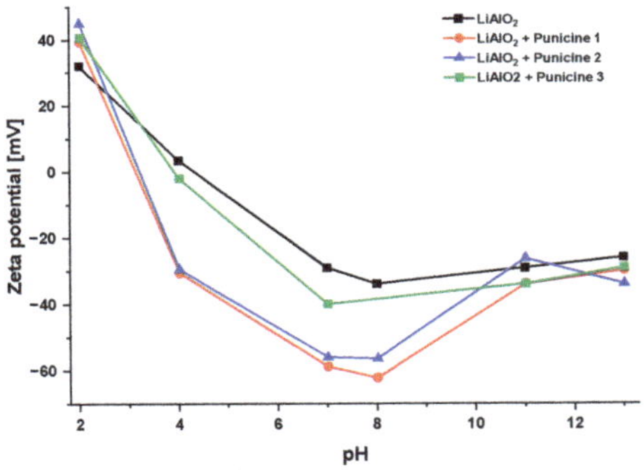

Figure 10. Zeta potential of pure lithium aluminate, lithium aluminate with 7.2 µL of 42.5 mmol/L punicine **1**, lithium aluminate with 7.2 µL of 42.5 mmol/L punicine **2**, and lithium aluminate with 7.2 µL of 42.5 mmol/L punicine **3** in distilled water and under different pH values.

The zeta potential of the gehlenite sample at its natural pH value in water was determined to be −23.53 mV. By adding 2.4 µL of punicine **3** (0.02 mM), the zeta potential increased to −9.18 mV. By increasing the concentration of punicine **3**, a gradual increase in the zeta potential of gehlenite was observed, with the zeta potential at a concentration of 0.04 mM being −0.49 mV, while a slight increase in the zeta potential was noted when the concentration was increased to 0.06 mM (+2.04 mV). The zeta potential of gehlenite at a pH of 2.0 ± 0.3 was also measured. The gehlenite used in our studies showed a zeta potential of +16.97 mV at a pH of 2.0 ± 0.3. After the addition of punicine **3** (0.02 mM), this increased to +37.93 mV and remained approximately in this range regardless of the added concentration of punicine **3** (see Supporting Information). This result can be attributed to at least two effects: As already shown in the blank experiment, the hydrophobicity of the side chain is an important factor. Secondly, the nucleophilicity of the oxygen atoms, which is higher in the deprotonated state, is apparently important. The substitution of at least one of the OH groups of $LiAl_2(OH)_7$ with a punicine molecule takes place, which leads to a considerable hydrophobization of the lithium aluminate surfaces.

The results shown in Section 3.3 show that oxidative plasma treatment leads to the partial dehydration of lithium aluminum hydroxide hydrate $LiAl_2(OH)_7 \times n\,H_2O$, which forms on the surface of lithium aluminate $LiAlO_2$ in water, and also when lithium aluminate is stored with access to moisture. IR spectra confirm these results (Figure 11).

Thus, plasma-treated $LiAlO_2$ mixed with 60 µL of the stock solution of punicine **3**, and non-plasma-treated $LiAlO_2$ mixed with the same amount of punicine **3** were measured. Without the plasma treatment of the sample, a large -OH peak at 3334 cm^{-1} was observed, in addition to a small peak due to small amounts of the collector on the surface. After a 5 min plasma treatment of the sample, the -OH peak had almost disappeared, thus indicating a hydrophobization of the lithium aluminum surface. This oxidative plasma-induced hydrophobization is reflected in the flotation results in the absence of punicines (cf. Figure 8, vide supra). The mere oxidative plasma treatment of lithium aluminate in these blind samples results in yield increases to 42.08 ± 1.18%, i.e., by 59%. On the other hand, this dehydration reduces the likelihood of the nucleophilic exchange of hydroxide for punicine molecules, as shown in Scheme 5. Here, the effects seem to balance each other out. Thus, the flotation yield with punicine **3** is almost identical under otherwise identical conditions after plasma pretreatment (81.29%; cf. Figure 9, vide supra) than without it (82.69%, cf. Figure 3, vide supra). For the further analysis of plasma-pretreated $LiAlO_2$ and its interaction with punicine **3** in the flotation process, zeta potential measurements were performed for various $LiAlO_2$ samples. Figure 12 shows the zeta potential of a pure $LiAlO_2$ sample not pretreated with plasma compared to a sample pretreated with oxidative plasma for 3 min. These are compared with analogous samples in the presence of 0.08 mM of punicine **3**. The influence of plasma on the surface charge is clearly seen. The zeta potential of pure $LiAlO_2$ without the plasma pretreatment was −31.80 mV. After a three-minute plasma treatment, the zeta potential increased to −18.16 mV, which is consistent with a less negatively charged surface due to the reduction in deprotonatable OH groups or the release of lithium cations from the lithium aluminum hydroxide hydrate, which is better soluble than $LiAlO_2$. This effect is reflected in the measurements in the presence of punicine **3**, where the value is less negative due to the substitution of OH groups of the $LiAl_2(OH)_7$ by punicines even before the plasma treatment. The plasma treatment then even leads to a change in sign (+3.67 mV) due to the presence of cationic pyridinium rings on the particle surface.

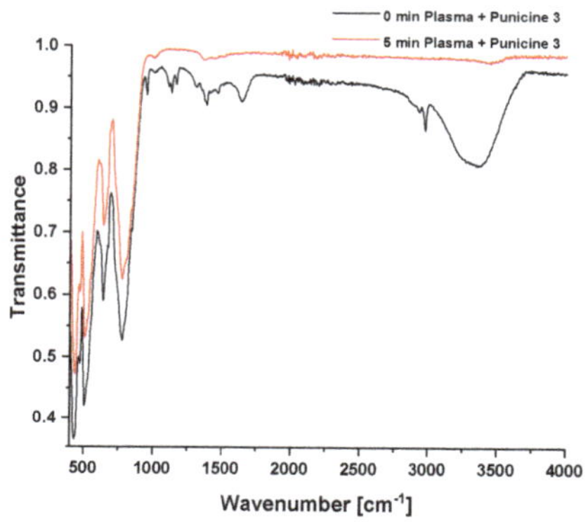

Figure 11. FTIR spectra of $LiAlO_2$ in the presence of 60 µL 42.5 µmol/L punicine **3** with and without pretreatment with plasma.

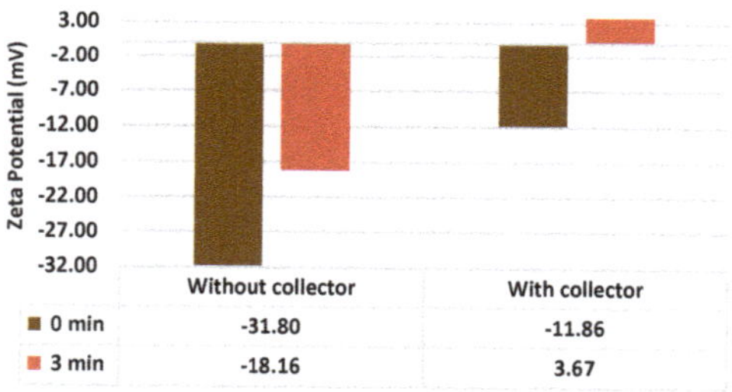

Figure 12. Zeta potential of pure and oxidative plasma-pretreated lithium aluminate, with and without 10 µL of 42.5 mmol/L punicine **3** at pH 10.9 ± 0.3.

Other effects are caused by reductive plasma, which results in significant yield losses, as described in Section 3.4.3. This effect can i.a. be attributed to a reduction in the surface area of $LiAlO_2$ which is literature-known [48]. The effect of the size reduction is due to the bombardment or etching effect of the reactive plasma species of Ar and H_2 [57]. This is already established as an advanced method for surface cleaning and/or thickness reduction in the Rapid Etching System (RES). Similar to the results we found, the literature also describes a one-minute treatment as the most electrochemically effective. It has been observed that further extensions of the exposure time (>1 min) did not result in any significant improvement [57]. When Ar/H_2 gas plasma is used, hydrogen gas is reduced to xH* [57,58]. The excited hydrogen atom interacts with the surface and causes a partial reduction on the surface of $LiAlO_2$, which, according to the flotation results, is then more hydrophilic than when treated with oxidative plasma. The reaction equation of this process was formulated as follows in the literature:

$$LiAlO_2 + xH^* \rightarrow LiAlO_{2-x}\frac{x}{2}H_2 + \frac{x}{2}O_2$$

One caveat should be mentioned here: the reactor was a mass reactor and did not behave like a fluidized bed reactor in which the solid particles behave like a liquid. Therefore, not all of the possible surface area of the $LiAlO_2$ sample in the plasma reactor chamber is exposed to the plasma effect. Nevertheless, the effects are significant.

5. Conclusions

Our studies show that the three derivatives of the natural product punicine, which have an adamantyl, 2-methylnonyl, and heptadecyl substituent at the 4-position of the pyridinium, can be used as collectors whose properties can be modulated by pH and light conditions. This allows the flotation parameters to be adjusted to optimize recovery rates or selectivity for the separation of unwanted gangue material. It was shown that the punicine with the long-chain heptadecyl group achieved the best recovery rates, thus demonstrating that there is a correlation between the hydrophilic index log P and the flotation results. α-Branching of the alkyl group has no effect on flotation. When using Hallimond tubes made of Pyrex glass for flotation processes, up to 153% better recovery rates of the engineered artificial mineral (EnAM) lithium aluminate ($LiAlO_2$) were achieved in the dark than in daylight. In all the cases, the flotation of the gangue material gehlenite ($Ca_2Al[AlSiO_7]$) was suppressed. The minimum recovery of this undesired gangue material was 16% after a one-minute plasma treatment followed by flotation at a pH of 8.0 ± 0.3 with heptadecyl punicine as a collector. In general, with the exception of gehlenite, plasma pretreatment has a significant influence on flotation results in blank tests and in the presence of punicines. While a one-minute plasma pretreatment increases the flotation results of lithium aluminate in blank tests by 58 to 42%, the yield of such pretreated samples can be increased to over 81%. As expected, the flotation result is influenced by several parameters. These include the chemical reaction of the punicines with the $LiAl_2(OH)_7$ on the lithium aluminate surface, with the substitution of OH groups for organic molecules. This, in turn, depends on how many OH groups are available and whether these have been changed by the oxidative plasma treatment. This intrinsically increases the hydrophobicity of the particle, while the probability of substitution by a punicine molecule decreases.

Supplementary Materials: The following supporting information can be downloaded at https://www.mdpi.com/article/10.3390/separations11120362/s1: Figure S1: ^{1}H-NMR 4-((1-adamantyl)methyl)pyridine; Figure S2: ^{13}C-NMR of 4-((1-adamantyl)methyl)pyridine; Figure S3: DEPT-135 of 4-((1-adamantyl)methyl)pyridine; Figure S4: ^{1}H-NMR 4-(2-methylnonyl)pyridine; Figure S5: ^{13}C-NMR 4-(2-methylnonyl)pyridine; Figure S6: DEPT-135 of 4-(2-methylnonyl)pyridine; Figure S7: ^{1}H-NMR of punicine 1; Figure S8: ^{13}C-NMR of punicine 1; Figure S9: DEPT-135 of punicine 1; Figure S10: ^{1}H-NMR of punicine 2; Figure S11: ^{13}C-NMR of punicine 2; Figure S12: DEPT-135 of punicine 2; Figure S13: ^{1}H-NMR of punicine 3; Figure S14: Zeta potential punicine **3** at different concentrations with $LiAlO_2$ and gehlenite at pH 7.0 ± 0.3; Figure S15: Zeta potential punicine **3** at different concentrations with $LiAlO_2$ and gehlenite at pH 2.0 ± 0.3; Figure S16: Zeta potential of punicine **3** at different concentrations and pH media with $LiAlO_2$; Figure S17: Single mineral flotation of lithium aluminate and gehlenite using punicine **2** as collector at pH 11 for lithium aluminate and pH 8.3 for gehlenite at different light scenarios. The conditioning was as follows: 120 μL collector, 1 min mixing, 30 μL frother, 1 min mixing, stirring speed 500 rpm, air flow rate 32 cm^3/min using 250 mL distilled water and flotation time 3 min; Figure S18: The figure shows the XRD of gehlenite; Figure S19: The figure shows the XRD of $LiAlO_2$; Figure S20: Illustration of the contact angle of water droplet on the surface of the different minerals' samples; Figure S21: Plasma setup; Figure S22. Dimensions of the Hallimond tube used. Pyrex glass; Figure S23: Dimensions of the box used; and Figure S24: Photograph of the box used.

Author Contributions: Conceptualization, A.S., A.Z. and M.H.F.; methodology, A.Z., M.H.F., U.E.A.F., A.P.W., A.W. and T.S.; validation, A.Z., M.H.F., S.M.T., I.E.H., A.W. and T.S.; investigation, A.Z., M.H.F.; S.M.T., I.E.H., S.Z., A.W. and T.S.; data curation, A.S., A.Z., A.W. and M.H.F.; writing—original draft preparation, A.Z. and A.S.; writing—review and editing, A.S.; supervision, A.S., A.P.W. and U.E.A.F.; project administration, A.S.; and funding acquisition, A.S. All authors have read and agreed to the published version of the manuscript.

Funding: This research was funded by the Deutsche Forschungsgemeinschaft DFG, grant number SCHM 1371/18-1, project no. 470324113 for A.S. within the priority program SPP 2315 "Engineered Artificial Minerals (EnAM)—a geo-metallurgical tool to recycle critical elements from waste streams". The design and construction of the plasma reactor used were carried out as part of the SFB 1366 (C2), Project-ID 394563137, and in the project with grant number WE 2331/21-2 as part of the Priority Program 2045, both funded by the Deutsche Forschungsgemeinschaft (DFG, German Research foundation).

Data Availability Statement: The original contributions presented in this study are included in the article/Supplementary Material. Further inquiries can be directed to the corresponding author(s).

Acknowledgments: We would like to thank Gundula Helsch and Thomas Peter for their help in the XRD and contact angle measurements. Special thanks to Jessica Streichert and Christian Rust for performing the TG-MS measurements.

Conflicts of Interest: The authors declare no conflicts of interest.

References

1. Kim, H.; Hong, J.; Park, N.Y.; Kim, S.W.; Kang, K. Aqueous rechargeable Li and Na ion batteries. *Chem. Rev.* **2014**, *144*, 788–827 [CrossRef]
2. Sousa, R.; Ramos, V.; Guedes, A.; de Sousa, A.B.; Noronha, F.; Leite, M.M. Flotation of lithium ores to obtain high-grade Li_2O concentrates. Are there any mineralogical limitations? *Int. J. Min. Mater. Metall. Eng.* **2019**, *5*, 2562–4571. [CrossRef]
3. Yu, W.H.; Zhou, J.H.; Hu, J.H.; Shang, Z.; Zhou, X.; Xu, S.M. Exploring the energy and environmental sustainability of advanced lithium-ion battery technologies. *Res. Conserv. Recycl.* **2025**, *212*, 107963. [CrossRef]
4. Zhou, L.M.; Zhang, K.; Hu, Z.; Tao, Z.; Mai, L.; Kang, Y.-M.; Chou, S.-L.; Chen, J. Recent developments on and prospects for electrode materials with hierarchical structures for lithium-ion batteries. *Adv. Energy Mater.* **2018**, *8*, 1701415. [CrossRef]
5. Ding, Y.L.; Cano, Z.P.; Yu, A.P.; Lu, Y.; Chen, Z. Automotive Li-ion batteries: Current status and future perspectives. *Electrochem. Energ. Rev.* **2019**, *2*, 1–28. [CrossRef]
6. Levich, B. *A Five Year Strategic Outlook for the Lithium Industry*; Metal Bulletin: London, UK, 2009.
7. U.S. Geological Survey. *Mineral Commodity Summaries 2024*; U.S. Geological Survey: Reston, VA, USA, 2024; p. 111. [CrossRef]
8. Kaunda, R.B. Potential environmental impacts of lithium mining. *J. Energy Nat. Resour. Law* **2020**, *38*, 237–244. [CrossRef]
9. Farjana, S.H.; Huda, N.; Mahmud, M.A.P. Life cycle assessment of cobalt extraction process. *J. Sustain. Min.* **2019**, *18*, 150–161. [CrossRef]
10. Kesler, S.E.; Gruber, P.W.; Medina, P.A.; Keoleian, G.A.; Everson, M.P.; Wallington, T.J. Global lithium resources: Relative importance of pegmatite, brine and other deposits. *Ore Geol. Rev.* **2012**, *48*, 55–69. [CrossRef]
11. Elwert, T.; Strauß, K.; Schirmer, T.; Goldmann, D. Phase Composition of High Lithium Slags from the Recycling of Lithium Ion Batteries. *World Metall.—ERZMETALL* **2012**, *65*, 163–171.
12. Qiu, H.; Li, H.; Fischlschweiger, M.; Ranneberg, M.; Graupner, T.; Lucas, H.; Stallmeister, C.; Friedrich, B.; Yagmurlu, B.; Goldmann, D. Valorization of lithium containing slags from pyrometallurgical recycling route of spent lithium-ion batteries: The enrichment of γ-$LiAlO_2$ phase from thermodynamic controlled and modified slags. *Miner. Eng.* **2024**, *217*, 108918. [CrossRef]
13. Li, H.; Qiu, H.; Schirmer, T.; Goldmann, D.; Fischlschweiger, M. Tailoring Lithium Aluminate Phases Based on Thermodynamics for an Increased Recycling Efficiency of Li-Ion Batteries. *ACS ES&T Eng.* **2022**, *2*, 1883–1895. [CrossRef]
14. Qiu, H.; Kersebaum, J.; Wollmann, A.; Feuge, N.; Haas, A.; Goldmann, D.; Wilhelm, R. Improvement of the froth flotation of $LiAlO_2$ and melilite solid solution via pre-functionalization. *Sci. Rep.* **2021**, *11*, 20443. [CrossRef]
15. Wu, H.; Tian, J.; Xu, L.; Fang, S.; Zhang, Z.; Chi, R. Flotation and adsorption of a new mixed anionic/cationic collector in the spodumene-feldspar system. *Miner. Eng.* **2018**, *127*, 42–47. [CrossRef]
16. Moon, K.S.; Fuerstenau, D.W. Surface crystal chemistry in selective flotation of spodumene ($LiAl[SiO_3]_2$) from other aluminosilicates. *Int. J. Miner. Process* **2003**, *72*, 11–24. [CrossRef]
17. Bulatovic, S.M. *Handbook of Flotation Reagents: Chemistry, Theory and Practice, Volume 3: Flotation of Industrial Minerals*; Elsevier: Amsterdam, The Netherlands, 2015; pp. 41–56.
18. Fuerstenau, M.C.; Jameson, G.; Yoon, R. *Froth Flotation a Century of Innovation*; Society for Mining, Metallurgy, and Exploration: Englewood, CO, USA, 2007.

19. Acker, S.; Namyslo, J.C.; Rudolph, M.; Strube, F.; Fittschen, U.E.A.; Qiu, H.; Goldmann, D.; Schmidt, A. Polyether-tethered imidazole-2-thiones, imidazole-2-selenones and imidazolium salts as collectors for the flotation of lithium aluminate and spodumene. *RSC Adv.* **2022**, *13*, 6593–6605. [CrossRef]
20. Hunter, T.N.; Pugh, R.J.; Franks, G.V.; Jameson, G.J. The role of particles in stabilizing foams and emulsions. *Adv. Colloid Interface Sci.* **2008**, *137*, 57–81. [CrossRef]
21. Polat, H.; Erdogan, D. Heavy metal removal from waste waters by ion flotation. *J. Hazard. Mater.* **2007**, *148*, 267–273. [CrossRef]
22. Rahman, R.M.; Ata, S.; Jameson, G. The effect of flotation variables on the recovery of different particle size fractions in the froth and the pulp. *J. Int. J. Miner. Process.* **2012**, *106*, 70–77. [CrossRef]
23. Zamboulis, D.; Pataroudi, S.; Zouboulis, A.; Matis, K. The application of sorptive flotation for the removal of metal ions. *Desalination* **2004**, *162*, 159–168. [CrossRef]
24. Filippov, L.; Farrokhpay, S.; Lyo, L.; Filippova, I. Spodumene flotation mechanism. *Minerals* **2019**, *9*, 372. [CrossRef]
25. Wang, Y.; Pan, Z.; Qin, W.; Jiao, F. A simple and rapid HPLC method for the quantitative determination of sodium oleate in flotation study. *Miner. Eng.* **2019**, *141*, 105842. [CrossRef]
26. Zgheib, A.; Acker, S.; Fischer, M.H.; Namyslo, J.C.; Strube, F.; Rudolph, M.; Fittschen, U.E.A.; Wollmann, A.; Weber, A.P.; Nieger, M.; et al. Lithium aluminate flotation by pH- and light-switchable collectors based on the natural product punicine. *RSC Adv.* **2024**, *14*, 9353–9364. [CrossRef]
27. Zgheib, A.; Fischer, M.H.; Namyslo, J.C.; Fittschen, U.E.A.; Wollmann, A.; Weber, A.; Schmidt, A. Photo-switchable collectors for the flotation of lithium aluminate for the recycling of the critical raw material lithium. *ChemSusChem* **2024**, *17*, e202301900. [CrossRef]
28. Nawwar, M.A.; Hussein, S.A.; Merfort, I. Leaf phenolics of Punica granatum. *Phytochemistry* **1994**, *37*, 1175–1177. [CrossRef]
29. Schmidt, A.; Mordhorst, T. Conjugated, cross-conjugated, and pseudo-cross-conjugated derivatives of a pyridinium alkaloid from Punica granatum. *Arkivoc* **2003**, *14*, 233–245. [CrossRef]
30. Schmidt, A.; Mordhorst, T.; Nieger, M. Investigation of a betainic alkaloid from Punica granatum. *Nat. Prod. Res.* **2005**, *19*, 541–546. [CrossRef]
31. Albrecht, M.; Schneider, O.; Schmidt, A. Redox active donor-substituted punicin derivatives. *Org. Biomol. Chem.* **2009**, *7*, 1445–1453. [CrossRef]
32. Schmidt, A.; Albrecht, M.; Mordhorst, T.; Topp, M.; Jeschke, G. Studies on photocatalytically active materials containing structure elements of a pyridinium alkaloid from Punica granatum. *J. Mater. Chem.* **2007**, *17*, 2793–2800. [CrossRef]
33. Albrecht, M.; Gjikaj, M.; Schmidt, A. Intermolecular interactions of punicin derivatives. *Tetrahedron* **2010**, *66*, 7149–7154. [CrossRef]
34. Schmidt, A.; Topp, M.; Mordhorst, T.; Schneider, O. Redoxactive derivatives of the betaine-alkaloid Punicin from Punica granatum. Synthesis and cyclovoltametry. *Tetrahedron* **2007**, *63*, 1842–1848. [CrossRef]
35. Otto, C.F.; Herzberger, C.; Liu, M.; Namyslo, J.C.; Nieger, M.; Freese, T.; Lederle, F.; Hübner, E.G.; Schmidt, A. Borane adducts of punicine and of its dehydroxy derivatives (pyridinium-1-yl)-2- and 3-phenolates. *Tetrahedron* **2020**, *76*, 131627. [CrossRef]
36. Steiner, F.; Zgheib, A.; Fischer, M.H.; Büttner, L.; Schmidt, A.; Breitung-Faes, S. In situ hydrophobization of lithium aluminate particles for flotations by dry grinding in the presence of punicines. *Minerals* **2024**, *14*, 650. [CrossRef]
37. Zhang, Z.; He, Q.; Zhang, X.; Yang, C. Photoredoxcatalysed regioselective synthesis of C-4-alkylated pyridines with N-(acyloxy)phthalimides. *Org. Biomol. Chem.* **2022**, *20*, 1969–1973. [CrossRef]
38. Liu, J.; Obando, D.; Liao, V.; Lifa, T.; Codd, R. The many faces of the adamantyl group in drug design. *Eur. J. Med. Chem.* **2011**, *46*, 1949–1963. [CrossRef]
39. Velick, S.F. The structure of the branched chain fatty acids in wool fat. An X-ray diffraction. *J. Am. Chem. Soc.* **1947**, *69*, 2317–2322. [CrossRef]
40. Wei, X.; Fankhänel, B.; Zhao, W.; Perminov, A.; Storti, E.; Liu, M.; Ma, G.; Fabrichnaya, O.; Charitos, A.; Volkova, O. The wettability of pure liquid aluminum on oxide substrates. *J. Am. Ceram. Soc.* **2024**, *108*, e20103. [CrossRef]
41. Kinoshita, K.; Sim, J.W.; Ackerman, J.P. Preparation and characterization of lithium aluminate. *Mater. Res. Bull.* **1978**, *13*, 445–455. [CrossRef]
42. Vanfleet, R.R.; Simmons, J.A.; Hill, D.W.; Chou, M.M.C.; Chai, B.H. Antiphase ordering and surface phases in lithium aluminate. *J. Appl. Phys.* **2008**, *104*, 093530. [CrossRef]
43. Kwon, S.W.; Kim, E.H.; Park, S.B. The formation of phase pure lithium aluminate from the bulky lithium and aluminum alkoxides. *J. Mater. Sci. Lett.* **1999**, *18*, 931–933. [CrossRef]
44. Poeppelmeier, K.R.; Kipp, D.O. Cation replacement in α-lithium aluminate. *Inorg. Chem.* **1988**, *27*, 766–767. [CrossRef]
45. Dronskowski, R. Reactivity and acidity of Li in lithium aluminum oxide ($LiAlO_2$) phases. *Inorg. Chem.* **1993**, *32*, 1–9. [CrossRef]
46. Available online: www.chemicalbook.com (accessed on 4 December 2024).
47. Thiel, J.P.; Chiang, C.K.; Poeppelmeier, K.R. Structure of lithium aluminum hydroxide dihydrate ($LiAl_2(OH)_7$ ▪ 2 H_2O). *Chem. Mater.* **1993**, *5*, 297–304. [CrossRef]
48. Evteeva, O.G.; Kotsupalo, N.P. Reaction of lithiumaluminate with water. *Izv. Sibirs. Otdel. Akad. Nauk SSSR. Ser. Khim. Nauk* **1969**, *6*, 70–75.
49. Kropachev, A.; Kalabskiy, I. Hydrometallurgical preparation of lithium aluminum carbonate hydroxide hydrate, $LiAl_4(CO_3)(OH)_{12}$ ▪ 3 H_2O from aluminate solution. *Miner. Eng.* **2020**, *155*, 106470. [CrossRef]

50. Nemeth, M.T.; Ford, R.B.; Taylor, T.A. Analysis of Mixtures of Gamma Lithium Aluminate, Lithium Aluminate Carbona Hydroxide Hydrate, and Lithium Carbonate. *MRS Online Proc. Libr.* **1997**, *496*, 159–165. [CrossRef]
51. Dutta, P.K.; Puri, M. Anion exchange in lithium aluminate hydroxides. *J. Phys. Chem.* **1989**, *93*, 376–381. [CrossRef]
52. Champenois, J.B.; Mesbah, A.; Dannoux-Papin, A.; Cau dit Coumes, C.; Dacheux, N. $LiAl_2(OH)_6OH.2\ H_2O$ solubility produ and dihydrogen radiolytic production rate under γ-irradiation. *J. Nucl. Mater.* **2018**, *508*, 92–99. [CrossRef]
53. Michel, A.E.; Usher, C.R.; Grassian, V.H. Heterogeneous and catalytic uptake of ozone on mineral oxides and dusts: A Knuds cell investigation. *Geophys. Res. Lett.* **2002**, *29*, 1665. [CrossRef]
54. Hirajima, T.; Mori, M.; Ichikawa, O.; Sasaki, K.; Miki, H.; Farahat, M.; Sawada, M. Selective flotation of chalcopyrite ar molybdenite with plasma pre-treatment. *Miner. Eng.* **2014**, *66–68*, 102–111. [CrossRef]
55. Jacewicz, D.; Wyrzykowski, D.; Żamojć, K.; Czerwińska, D.; Czaja, P.; Chmurzyński, L. Thermal Properties of Potassiu Bis(Oxalato)Diaquochromates(III) in Solid State. Trans–Cis Isomerization of the $[Cr(C_2O_4)_2(OH_2)_2]^-$ Complex Ion in Aqueo Solutions. *Struct. Chem.* **2011**, *23*, 333–340. [CrossRef]
56. Shaban, M.; Merkert, N.; van Duin, A.C.T.; van Duin, D.; Weber, A.P. Advancing DBD Plasma Chemistry: Insights into Reacti Nitrogen Species such as NO_2, N_2O_5, and N_2O Optimization and Species Reactivity through Experiments and MD Simulatic *Environ. Sci. Technol.* **2024**, *58*, 16087–16099. [CrossRef] [PubMed]
57. Lan, C.; Chuang, S.I.; Bao, Q.; Liao, Y.-T.; Du, J.G. One-step argon/nitrogen binary plasma jet irradiation of $Li_4Ti_5O_{12}$ for stab high-rate lithium ion battery anodes. *J. Power Sources* **2015**, *275*, 660–667. [CrossRef]
58. Dessemond, C.; Soucy, G.; Laroche, N. Reductive thermal plasma conversion of a spodumene concentrate and impact on lithiu extraction. *Miner. Eng.* **2024**, *211*, 108682. [CrossRef]

separations MDPI

Article

Extraction and Counter-Current Separation of Phenylpropanoid Glycosides from *Pedicularis oederi* Vahl by Deep Eutectic Solvent

Yao Wang [1,†], Jun Dai [2,†], Xiaoting Zhang [1], Yuhan Wang [1], Fangfang He [1], Lu Liang [1], Duojie Longzhu [1,*] and Denglang Zou [1,*]

1 School of Life Science, Qinghai Normal University, Xining 810000, China; wwwyao9@163.com (Y.W.); zxt11242024@163.com (X.Z.); wx28465@sina.com (Y.W.); 15859557623@163.com (F.H.); ll13579100@outlook.com (L.L.)
2 Engineering Technology Research Center of Plant Cell Engineering, West AnHui University, Lu'an 237000, China; wxxydaijun@163.com
* Correspondence: longguo.2009@163.com (D.L.); dlangzou@foxmail.com (D.Z.)
† These authors contributed equally to this work.

Abstract: Deep eutectic solvents (DESs) are mixtures of organic compounds displaying excellent solvent properties while keeping an ecofriendly character. In this study, DESs have been applied to the extraction of phenylpropanoid glycosides from *Pedicularis oederi* Vahl, successively separated by means of counter-current chromatography. Firstly, the ultrasonic-assisted extraction conditions were optimized by response surface methodology, and the results showed phenylpropanoid glycosides could be well extracted under the optimized extraction conditions with deep eutectic solvents. Then, the sample was separated by counter-current chromatography using ethyl acetate/aqueous solution of choline chloride and glycerol (6:6, v/v) as the solvent system. In about 360 min, four phenylpropanoid glycosides, including 31.6 mg of echinacoside, 65.3 mg of Jionoside A1, 28.9 mg of Forsythoside B, 74.1 mg of verbascoside, and 21.2 mg of kaempferol-3-*O*-rutinoside were obtained from about 900 mg of the sample. It revealed deep eutectic solvents could be well employed as a green solvent for the extraction and counter-current separation of natural products.

Keywords: deep eutectic solvent; extraction; counter-current separation; response surface methodology; phenylpropanoid glycosides; *Pedicularis oederi* Vahl

1. Introduction

In the realm of traditional Tibetan medicinal practices, the herb known as *Pedicularis oederi* Vahl occupies a prominent position, predominantly thriving in the elevated terrains of Tibet, along with the provinces of Qinghai, Sichuan, and Gansu within China [1]. Research has validated its efficacy in addressing numerous forms of ailments, such as hepatic inflammation, gallbladder inflammation, edema, spermatorrhea, and ear ringing [2]. Phenylpropanoid glycosides, the primary bioactive compounds identified within this plant, exhibit a spectrum of pharmacological effects. These effects span antihepatotoxic, anti-inflammatory, antibacterial, analgesic, oxidation inhibitory, and genotoxicity prevention capabilities [3]. Given the broad spectrum of its pharmacological benefits, there exists an acute demand to isolate and purify these compounds' substantial volumes to serve as reference materials for analytical purposes and to facilitate in-depth pharmacological research.

Deep eutectic solvents (DESs), a type of ambient temperature liquid salt mixture formed by combining a hydrogen bond donor and a hydrogen bond acceptor, have been widely utilized in extracting bioactive compounds from various plant sources [4,5]. These compounds include anthocyanins extracted from raspberries, flavonoids isolated from Pollen Typhae, phenolic compounds from grape skins, and significant flavonoids from Oroxylum indicum seeds, as documented in several studies [6–8]. DESs offer numerous advantages, such as non-volatility, environmental friendliness, cost efficiency, and versatility, making them attractive for various green chemistry applications. However, challenges like high viscosity, limited thermal stability, and the need for specific handling conditions highlight the areas where further research and development are needed [9,10].

Traditional methods of column chromatography have been widely employed for extracting and purifying compounds from botanical sources. Despite their widespread use, these methodologies are known for being labor-intensive, slow, and inefficient in terms of solvent use, often necessitating multiple stages to complete. Moreover, they are prone to causing the permanent adsorption of substances onto the column's solid phase [11–13]. In contrast, preparative HPLC and high-speed counter-current chromatography (HSCCC) stand out as superior techniques, offering significant benefits such as enhanced efficiency, superior resolution, and consistent results [14–16]. These methods leverage advanced separation capabilities, inline monitoring, and automated management to facilitate the effective isolation of desired molecules [17–19]. Nonetheless, preparative HPLC typically does not accommodate crude extracts directly. Distinguished by its support-free liquid–liquid partitioning mechanism, HSCCC avoids the drawbacks of solid phase sample adsorption inherent in traditional chromatography, thereby minimizing the risk of sample degradation [20–23]. This method is particularly adept at extracting components from unrefined extracts without the need for prior processing, allowing for substantial sample volumes [12,22,24,25]. Consequently, HSCCC is gaining recognition for its utility in isolating pharmaceutical agents, particularly in the preparative extraction of active elements from natural product extracts. DESs present a sustainable, versatile, and cost-effective alternative to traditional solvents, aligning well with green chemistry principles. Their biodegradability, low toxicity, and ease of customization make them a promising choice for various industrial, pharmaceutical, and environmental applications. Considering the above advantages and previous reports, a DES based on choline chloride/glycerol was prepared for subsequent extraction and HSCCC separation of these bioactive phenylpropanoid glycosides from *Pedicularis oederi* Vahl.

In this study, a DES was employed as a green solvent to fuel the extraction and counter-current separation of phenylpropanoid glycosides from *Pedicularis oederi* Vahl to pave the way for further medicinal applications. Employing a minimalistic approach to resource usage, response surface methodology (RSM) utilizes quantitative data from tailored experimental designs to address and resolve complex multivariate equations effectively [26]. Its application has been extensively documented for the optimization of operational conditions [27–29]. As such, this study has applied RSM to refine the conditions under which ultrasonic-assisted extraction is performed. Eventually, four phenylpropanoid glycosides combined with a flavonoid were acquired, and their configurations were elucidated through ^{1}H-NMR analysis. This strategy heralds the current frontier in the DES application of natural product extraction and separation.

2. Materials and Methods

2.1. Apparatus

The HSCCC separation was performed employing a TBE-1000A system from Shanghai Tauto Biotech, Co., Ltd. (Shanghai, China). This system was notably outfitted with a trio of PTFE preparative coils, collectively encompassing a volume of 1000 mL, alongside an 80 mL sample loop. Further enhancements to the system included the incorporation of a TBP-5002 model constant-flow pump (Shanghai, China), a model sensing unit module of UV500 operational at 280 nm (Shanghai, China), and a N2000 model workstation (Shanghai, China), the latter being a product of Zhejiang University, Hangzhou, China. Temperature regulation during the experiments was achieved through a constant-temperature circulating device DC-0506, courtesy of Shanghai Sunny Hengping Scientific Instruments Co., Ltd. (Shanghai, China). Moreover, the analysis via a UPLC was performed via a Waters Acquity UHPLC I-Class system from Waters Corporation (Milford, CT, USA), which was distinguished by its binary solvent management system and an automated sample handling feature.

2.2. Reagents and Plant Material

For the extraction and separation processes, we exclusively utilized solvents of analytical grade, sourced from Macklin (Shanghai, China). The methanol with chromatographic purity for UPLC analysis was purchased from Yuwang Chemical Ltd. (Shandong, China). Throughout our experimentation, we employed deionized water. Pedicularis oederi Vahl, the plant material under study, was gathered from the Maixiu region within Qinghai Province, China, and subsequently reduced to a fine powder, achieving a granularity between 20 and 40 mesh prior to the extraction process. Furthermore, we obtained choline chloride (ChCl, purity of 99% or higher) and glycerol (purity of 99% or higher) from Aladdin (Shanghai, China).

2.3. UPLC Analysis

In this study, chromatographic separation was achieved utilizing the Acquity UPLC HSS T3 Column with dimensions of 2.1 × 100 mm and a 1.8 μm particle size, provided by Waters, headquartered in Milford, USA. The separation temperature was maintained at 35 °C. A binary mobile phase system was utilized, comprising water with 0.1% formic acid (A) and acetonitrile (B). A gradient elution method was applied, initiating with 5% of component B, which was linearly increased to 100% over 8 min, followed by a constant 100% of component B for an additional 2 min. The flow was set at 0.3 mL/min, and the quantity of each sample injected was precisely 1.0 μL. The analysis was conducted with the column oven set to a temperature of 35 °C, and absorbance was monitored at a wavelength of 320 nm.

2.4. Preparation of DES

Following the previously outlined procedure, the components required for the hydrogen bond interaction, namely ChCl as the acceptor and glycerol as the donor, were precisely measured and combined. Utilizing a magnetic stirrer, the mixture with a ratio of 1:2 was heated and agitated at a temperature of 80 °C. This process was continued until the resultant mixture achieved a transparent and uniform consistency. To preserve its quality until further application, the freshly synthesized deep eutectic solvent (DES) was subsequently placed in a desiccator for storage.

2.5. Optimization of Ultrasonic-Assisted Extraction Conditions

Response surface methodology (RSM) was employed to enhance the conditions for the ultrasonic-assisted extraction (UAE) of the target compounds. In this study, the variables considered were ultrasonic power (X1), the water content of deep eutectic solvents (DESs) (X2), and the ratio of liquids to solids (X3). The dependent variable, represented by Y, was characterized as the proportion of the maximum region for target compounds. A total of 17 experimental runs were designed, comprising 12 factorial points and five central point replicates (Table 1). For an investigation of the correlation between independent and dependent variables, the methodology of multiple regression based on least squares was employed. A robust model is further confirmed by substantial values in adjusted and predicted R^2 values, verifying its fit to the dataset [25]. The study of factor interactions was visualized through surface plots based on the selected model.

Table 1. Box–Behnken design matrix and experimental response.

Std	ID	Run	Factor 1	Factor 2	Factor 3	Response 1
			A: Ultrasonic Power	**B: Water Content**	**C: Liquid/Solid Ratio**	**Content of Targets**
			W	%	1	%
17	13	1	300	40	20	63.41
15	13	2	300	40	20	66.06
5	5	3	200	40	10	55.86

Table 1. *Cont.*

Std	ID	Run	Factor 1	Factor 2	Factor 3	Response 1
			A: Ultrasonic Power	B: Water Content	C: Liquid/Solid Ratio	Content of Targets
13	13	4	300	40	20	65.99
8	8	5	400	40	30	60.15
7	7	6	200	40	30	60.52
12	12	7	300	50	30	65.7
6	6	8	400	40	10	52.58
1	1	9	200	30	20	50.23
16	13	10	300	40	20	64.85
10	10	11	300	50	10	58.33
11	11	12	300	30	30	52.88
3	3	13	200	50	20	69.44
2	2	14	400	30	20	54.58
14	13	15	300	40	20	64.52
4	4	16	400	50	20	60.97
9	9	17	300	30	10	50.15

2.6. Ultrasonic-Assisted Extraction

One kilogram of *Pedicularis oederi* Vahl, after drying, was finely pulverized before being subjected to a trio of ultrasonic-assisted extraction (UAE) processes. These extractions were performed utilizing a deep eutectic solvent (DES) formulation that included a 50% water mixture while operating at a 230-watt power level and upholding a 24 liquid-to-solid ratio. The combined extracts, post-extraction, were then subjected to a vacuum drying process, yielding a crude extract weighing 171 g. This extract underwent further processing by being introduced into a D101 resin column, sized at 2.5 cm by 60 cm, for purification. Water was employed to wash away the DES from the extract. Subsequently, the fraction representing 80% of this purified extract was earmarked for further refinement via HSCCC.

2.7. Selection of Two-Phase Solvent System

When choosing the suitable solvent system, the choice was determined by the partition coefficient (***K***) associated with the sample's key constituents. This coefficient was evaluated through the HPLC methodology. Initially, a moderate quantity of the crude extract was dissolved in a solvent system consisting of two phases that had been pre-equilibrated. This mixture was then transferred to a separation funnel. Following a vigorous shake to ensure complete mixing of the two phases, 2 mL from each phase was removed and allowed to evaporate until dry. The resultant dry residues were prepared via methanol for subsequent examination via UPLC. The calculation of the ***K*** value involved comparing the upper phase's peak area to that in the lower phase, represented as a ratio.

2.8. Preparation of Two-Phase Solvent System and Sample Solution

In the method described, a ternary solvent mixture comprising ethyl acetate/aqueous solution of choline chloride and glycerol with a volume ratio of 6:6 was utilized. This blend was placed into a separation funnel and allowed to attain equilibrium at room temperature. Following equilibrium, the clearly defined upper and lower stages were meticulously separated and underwent a 30 min degassing procedure in an ultrasonic bath, conducted just before their application. For the HSCCC separation process, the preparation of the sample solution entailed dissolving 1.0 g of the crude extract's dry powder in 80 mL of the solvent system's upper phase.

2.9. High-Speed Counter-Current Chromatography Separation Procedure

To commence each trial, HSCCC was initially flushed with ethanol to eliminate any residual substances. Subsequently, the lower phase in the stationary stage was introduced to completely occupy the multilayer column. The lower phase was employed as a stationary phase with flow rate of 10 mL/min. The revolution speed was set at 1100 rpm with retention of the stationary phase at 65%, and the sample loading amount was 1.0 g. The separation temperature was set at 45 °C with a detection wavelength of 320 nm.

Once the leading edge of the solvent appeared and a stable hydrodynamic balance was achieved within the system, the sample was administered into HSCCC via the sample injection port. This was immediately followed by data acquisition. The divided fractions were gathered manually according to the chromatogram's instructions and then dehydrated using a vacuum. Following that, the desiccated specimens were reestablished in methanol for a subsequent assessment of their purity via UPLC analysis.

3. Results and Discussion

3.1. Optimization of Ultrasonic-Assisted Extraction Conditions

In the designed study, we conducted 17 trials in pairs, presenting the outcomes in Table 1. The R^2, R^2-adj, and R^2-predicted values stood at 98.64%, 96.90%, and 90.05%, respectively. These statistics highlight the superior efficacy of full quadratic models over alternative approaches in predicting Y.

To assess the impact of various factors on chlorogenic acid levels, the analysis of variance (ANOVA) methodology was employed, with the findings detailed in Table 2. The significance of each model component was inferred from the F-value's magnitude and the reciprocal relationship of the p-value, as suggested in [26]. Consequently, significant influences on chlorogenic acid content were observed for the linear components (X_1, X_2, and X_3), the interaction term ($X_{1\times2}$), and the quadratic components ($X_1{}^2$, $X_2{}^2$, and $X_3{}^2$), all registering p-values below the 0.05 threshold. In contrast, the interaction components ($X_{1\times3}$ and $X_{2\times3}$) were found to be statistically insignificant, with p-values exceeding 0.05. The model's precision in forecasting was further validated through an absence of compatibility analysis, which yielded a significance level exceeding 0.05, thus affirming the model's reliability in forecasting variations, as cited from [30].

Table 2. Analysis of variance.

Source	Sum of Squares	df	Mean Square	F-Value	p-Value	
Model	589.59	9	65.51	56.48	<0.0001	significant
A—Ultrasonic power	7.55	1	7.55	6.51	0.0381	
B—Water content	271.45	1	271.45	234.04	<0.0001	
C—Liquid/solid ratio	62.33	1	62.33	53.74	0.0002	
AB	41.09	1	41.09	35.43	0.0006	
AC	2.12	1	2.12	1.83	0.2187	
BC	5.38	1	5.38	4.64	0.0682	
A^2	33.58	1	33.58	28.96	0.0010	
B^2	46.88	1	46.88	40.42	0.0004	
C^2	99.62	1	99.62	85.90	<0.0001	
Residual	8.12	7	1.16			
Lack of fit	3.24	3	1.08	0.8854	0.5207	not significant
Pure error	4.88	4	1.22			
Cor total	597.71	16				

In three-dimensional graphical representations, regression analyses are vividly portrayed, enabling an exploration of the dynamics between experimental variables and their resultant responses, as well as interactions among pairs of variables being tested. This method showcases how independent variables correlate with dependent variables through three-dimensional surface plots of response for Y (Figure 1). Within these plots, two variables are visualized simultaneously on a single three-dimensional surface, with a third variable maintained at a baseline level of zero.

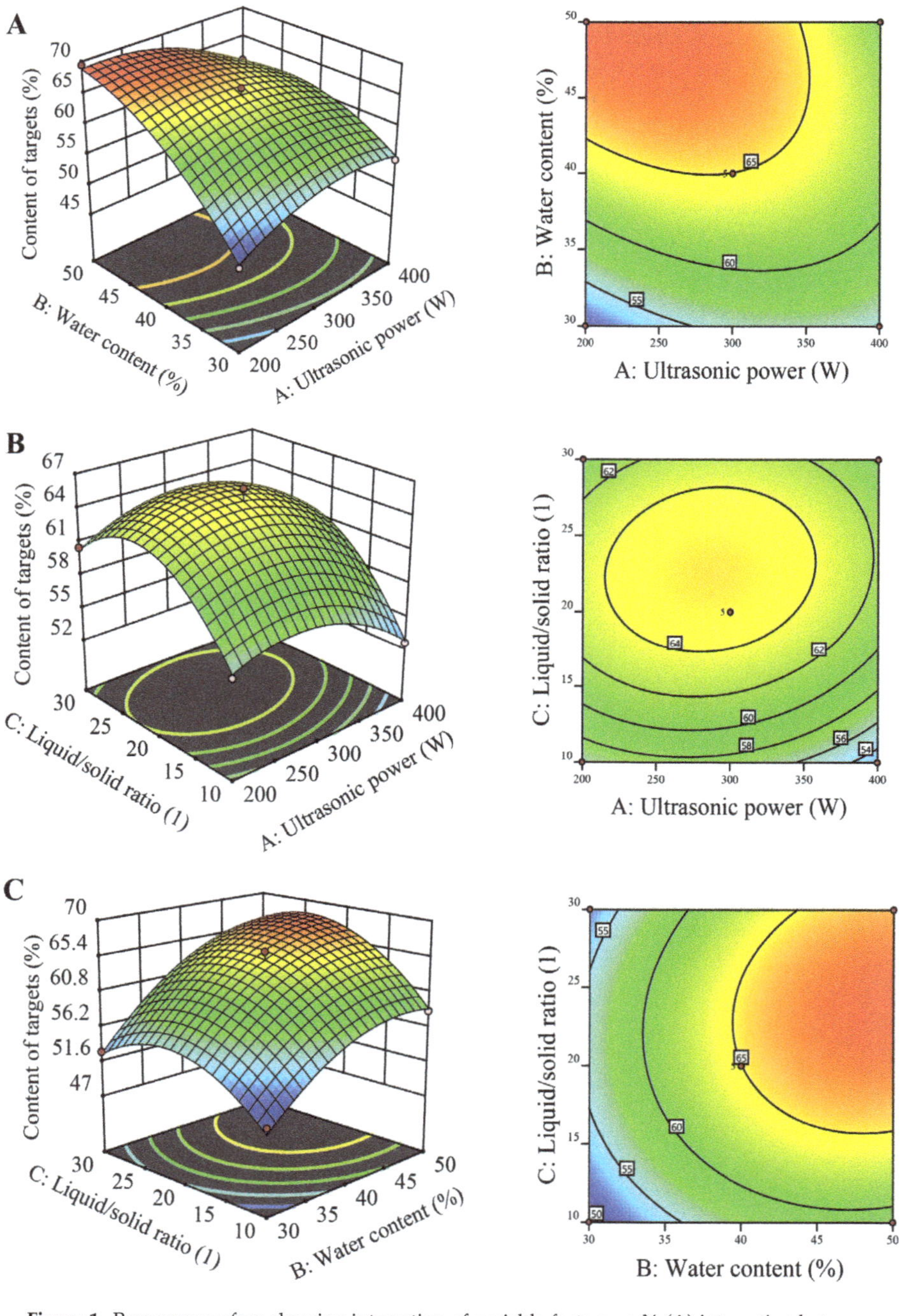

Figure 1. Response surface showing interaction of variable factors on Y: (**A**) interaction between ultrasonic power and water content; (**B**) interaction between ultrasonic power with liquid/solid ratio; (**C**) interaction between water content and liquid/solid ratio.

Figure 1A showed the interaction between ultrasonic power and water content. Initially, *Y* increased by increasing the water content, and then *Y* reached a constant level. Based on an excellent extraction ability toward phenylpropanoid glycosides, the extraction and enrichment of compounds within the optimal water content range for DESs have demonstrated their potential to be as effective as traditional methods such as liquid–liquid extraction or chromatography using macroporous resins. Initially, an increase in water content resulted in an increase in the yield (***Y***), demonstrating the efficiency of this process. However, with the increase in water, the content of phenylpropanoid glycosides maintained a constant level due to the fact that most of the phenylpropanoid glycosides had been extracted, and it had a tiny effect on the extraction procedure for further increases in the water content in DESs. The effectiveness of DESs in isolating substances from organic origins has been shown to be remarkably effective.

In Figure 1B, the correlation between the liquid/solid ratio and ultrasonic power is depicted. At first, a rise in ***Y*** was noted as the ratio of liquids to solids increased, but this trend reversed beyond a certain point. This behavior can be exemplified by the diverse polarity range found in the crude extract of *Pedicularis oederi* Vahl. As the liquid-to-solid ratio increased, both the quantities of phenylpropanoid glycosides and additional constituents experienced a simultaneous escalation. However, upon surpassing a liquid-to-solid ratio of 24, the growth rate of the sum of concentrations of other compounds ($\sum C_j$) surpassed that of the target compounds ($\sum C_i$), leading to a reduction in ***Y***.

The optimal parameters for the ultrasonic-assisted extraction (UAE) procedure were determined to be a sonic energy input of 230.61 W, a 50% water content in deep eutectic solvents (DESs), and a liquid-to-solid ratio of 23.54, predicting a *Y* of 69.60%. Adapting to real-world conditions, the settings were adjusted to a 230 W power with a 50% water content in DESs and a liquid-to-solid ratio of 24. Performed three times under these ideal circumstances, the experiment yielded a *Y* of 68.52%. It affirms the efficacy of the RSM, coupled with a well-considered experimental design, in fine-tuning the UAE process for the enhanced extraction and concentration of desired compounds. These findings endorse UAE as a favorable method for compound extraction and enrichment.

3.2. Selection of High-Speed Counter-Current Chromatography Experimental Conditions

In initiating an HSCCC separation procedure, selecting a suitable solvent system is paramount to achieving the preferred ***K*** value for the substances of interest. This selection process is largely influenced by the compounds' chemical properties. Various critical factors, including the polarity of the sample (evaluated based on partition coefficient values), its dissolution rate, presence of ions, and capacity to create complexes, must be carefully considered. Ideally, the partition coefficient (K) must fall within the 0.2 to 5 range [21,31,32], and for a successful separation of two compounds, the separation factor ($\alpha = K_2/K_1$, where $K_2 > K_1$) needs to be above 1.5 [21,33]. Solute elution close to the solvent front, resulting from a significantly low *K* value, leads to reduced resolution, whereas a high *K* value can enhance resolution but at the cost of producing broader and more diluted peaks due to extended elution times [31,34–36]. It is critical that the compounds under study remain steady and easily dissolved, with the solvent system capable of rapidly and neatly dividing into two separate phases.

To optimize the HSCCC separation, we systematically assessed various biphasic solvent combinations comprising ethyl acetate/aqueous solution of choline chloride and glycerol in varying ratios (5:9, 5:8, 5:7, 5:6, *v*/*v*/*v*), evaluating the *K* values for the desired substances and consolidating the results in Table 3. In the beginning, the solvent system of ethyl acetate/aqueous solution of choline chloride and glycerol with a 5:9 proportion (*v*/*v*/*v*) was examined for its effectiveness in distributing phenylpropanoid glycosides. These glycosides were predominantly found in the lower phase, indicating an excessively high polarity for this system in comparison to chlorogenic acid. Subsequently, adjustments were made to the DES ratio to lower the system's polarity. While such modifications allowed for the phenylpropanoid glycosides to be distributed in the upper phase, achieving the optimal partition coefficient range ($0.2 < \boldsymbol{K} < 5$) for these glycosides remained elusive. As a result, we raised the ratio of ethyl acetate, which effectively adjusted the distribution of the phenylpropanoid glycosides. The final selection of the ethyl acetate/aqueous solution of choline chloride and glycerol at a 6:6 ratio (v/v) provided satisfactory K and α values, aligning with the established criteria for effective separation.

Table 3. K values of target compounds.

Solvent Systems	K_2	K_3	K_4	K_5
5:9 (4:5)	0.14	0.08	0.16	0.21
5:8 (3:5)	0.20	0.19	0.26	0.32
5:7 (2:5)	0.33	0.41	0.55	0.62
5:6 (1:5)	0.42	0.69	0.78	0.93
6:6 (1:5)	0.57	0.90	1.39	2.17
7:6 (1:5)	0.94	1.12	1.49	2.66

Exploring various parameters beyond solvent configurations, this study delved into the impacts of mobile phase flow rates and apparatus rotation speeds on chromatographic outcomes. The mobile phase was delivered at varying rates (8.0, 10.0, and 12.0 mL/min) to assess its influence on both the duration of separation and the clarity of chromatographic peaks. It was observed that decreased flow rates extended the separation process yet enhanced peak clarity, whereas increased rates resulted in reduced separation quality. Consequently, a moderate flow rate of 10.0 mL/min was selected to proceed with HSCCC experiments. Furthermore, the rotation speed of the device was found to significantly affect the preservation of the immobile stage, with higher speeds leading to potential emulsification issues. Therefore, an optimal speed of 1100 revolutions per minute (rpm) was established, achieving a 65% stationary phase retention. Interestingly, this setup also facilitated the isolation of a flavonoid compound with satisfactory purity using a reverse rotation approach in the HSCCC process.

Under optimized parameters, the four phenylpropanoid glycosides combined with a flavonoid were obtained in approximately 360 min for a single iteration, as depicted in Figure 2. The five targets were subjected to NMR for a chemical structure analysis. Eventually, four phenylpropanoid glycosides, including 31.6 mg of echinacoside, 65.3 mg of Jionoside A1, 28.9 mg of Forsythoside B, 74.1 mg of verbascoside, and 21.2 mg of kaempferol-3-*O*-rutinoside were obtained from 900 mg of the sample. Analysis via UPLC indicated that the final isolated compounds possessed a purity exceeding 85.23%, as depicted in Figure 3. The above results indicated HSCCC offers an efficient, scalable, and gentle method for the separation and purification of phenylpropanoid glycosides, preserving their structural integrity and bioactivity.

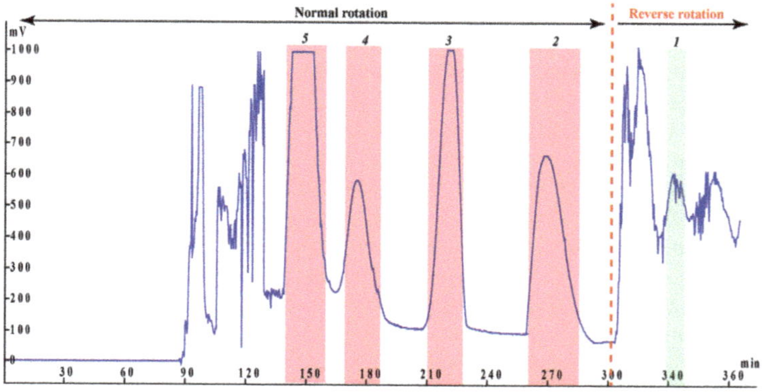

Figure 2. High-speed counter-current chromatography (HSCCC) chromatogram of the sample using the ethyl acetate/aqueous solution of choline chloride and glycerol (6:6, v/v). Conditions: stationary phase, lower phase; flow rate, 10 mL/min; revolution speed, 1100 rpm; sample amount, 1.0 g; separation temperature, 45 °C; detection wavelength, 320 nm; retention of the stationary phase: 65%.

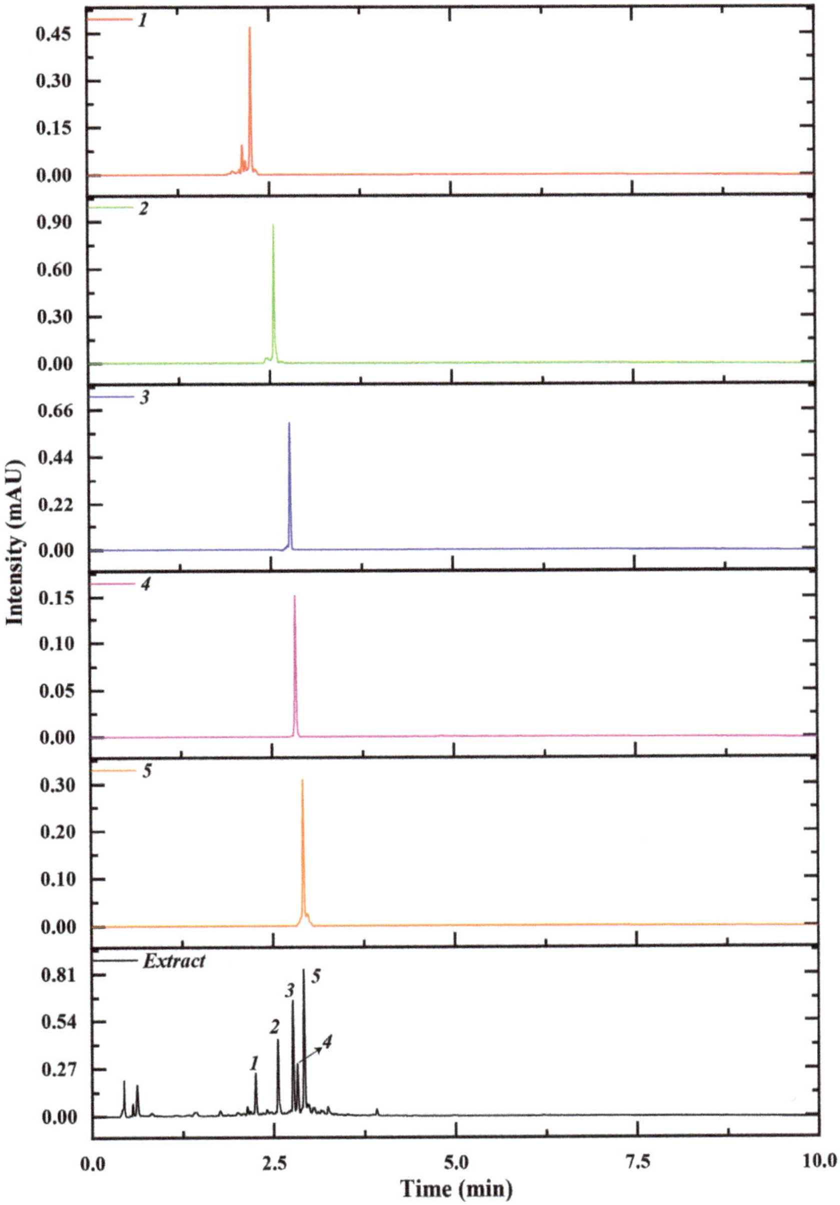

Figure 3. UPLC chromatograms of the fractions obtained through HSCCC, 80% MR elution fraction, and the crude extract. Conditions: Separation was carried out using an Acquity UHPLCHSS T3 column (2.1 × 100 mm, 1.8 μm; Waters) at a maintained temperature of 35 °C. The delineated mobile phases included water supplemented with 0.1% formic acid (A) and acetonitrile (B). The elution gradient was as follows: 5–100% B in 0–8 min and 100% B in 8–10 min, operating at a flow rate of 0.3 mL/min. The procedure involved an injection volume set at 1.0 μL. Throughout the process, a consistent column temperature of 35 °C was upheld, coupled with a detection wavelength pinpointed at 320 nm.

3.3. Structural Identification

The target compounds' chemical structures were ascertained using 1H-NMR analysis (Figure 4), with the results summarized as follows:

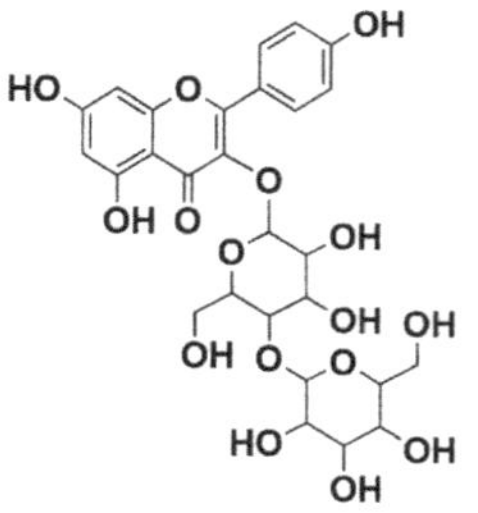

1

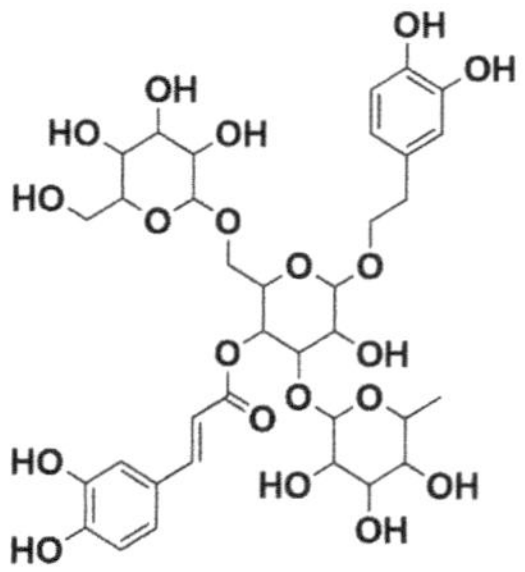

2

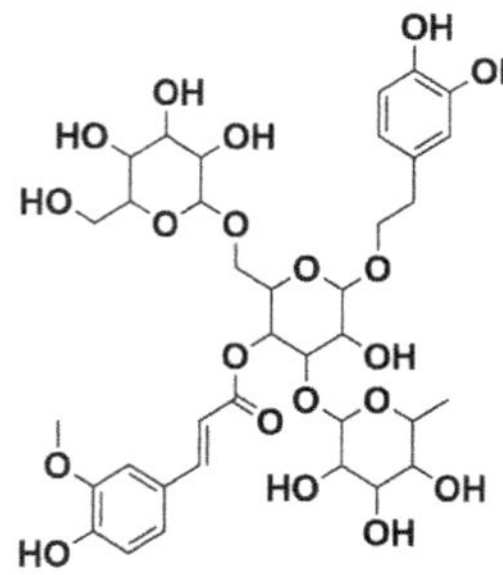

3

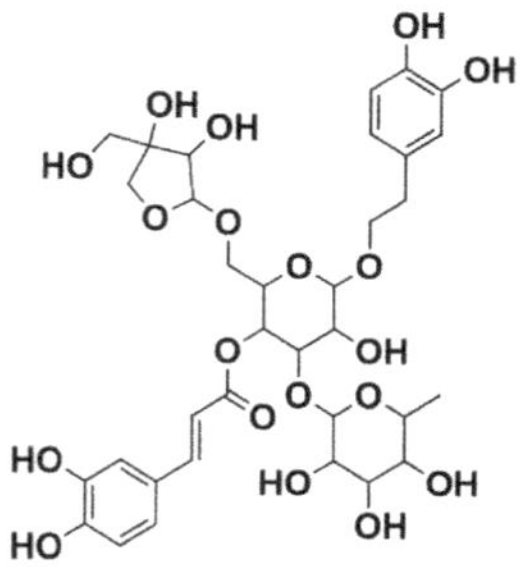

4

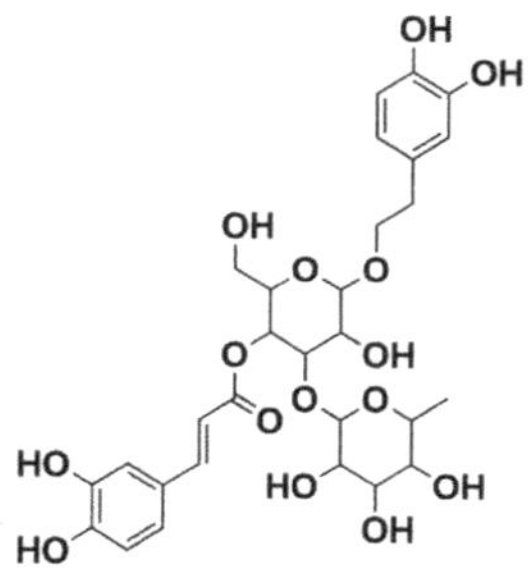

5

Figure 4. The chemical structures of kaempferol-3-O-rutinoside (Target **1**), echinacoside (Target **2**), Jionoside A1 (Target **3**), Forsythoside B (Target **4**), and verbascoside (Target **5**).

Target **1**: HRMS (m/z) 595.1607 $[M + H]^+$ (calcd. for C27H30O15, 595.1657). ^{1}H-NMR (DMSO-d_6, 600 MHz) δ: 8.01 (2H, d, J = 8.7 Hz, 2′-H and 6′-H), 6.87 (2H, d, J = 8.6 Hz, 3′-H and 5′-H), 6.41 (1H, d, J = 1.9 Hz, 8-H), 6.22 (1H, d, J = 1.8 Hz, 6-H), 5.31 (1H, d, J = 7.4 Hz, 1″-H), 4.36 (1H, bs, 1‴-H), 3.69 (1H, bd, J = 10.1 Hz, 6a″-H), 3.69 (2H, m, 3‴-H and 5‴-H), 3.31 (1H, bd, J = 10.2 Hz, 6b″-H), 3.25 (2H, m, 5″-H and 2‴-H), 3.20 (1H, m, 3″-H), 3.15 (1H, m, 2″-H), 3.11 (1H, m, 4‴-H), 3.07 (1H, m, 4″-H), 0.98 (3H, d, J = 6.1 Hz, 6‴-H). The structure was determined as kaempferol-3-O-rutinoside based on the results of ^{1}H NMR [37].

Target **2**: HRMS (m/z) 804.2845 $[M + NH_4]^+$ (calcd. for C35H46O20, 804.2920). ^{1}H-NMR (DMSO-d_6, 600 MHz) δ (ppm): 7.48 (d, 1H, J = 14.9 Hz, H-7′), 7.03 (d, 1H, J = 1.8 Hz, H-2′), 6.99 (dd, 1H, J = 2.1 Hz, J = 8.8 Hz, H-6′), 6.78 (d, 1H, J = 8.3 Hz, H-5′), 6.66 (d, 1H, J = 1.9 Hz, H-2), 6.65 (d, 1H, J = 8.1 Hz, H-5), 6.52 (dd, 1H, J = 2.1 Hz, J = 8.4 Hz, H-6), 6.15 (d, 1H, J = 15.7 Hz, H-8′), 5.02 (s, 1H, Rha H-1), 4.72 (t, 1H, J = 9.7 Hz, Glu H-4), 4.35 (d, 1H, J = 7.91 Hz, Glu′ H-1), 4.14 (d, 1H, J = 7.82 Hz, Glu H-1), 3.88 (m, 1H, H-8), 3.71–2.92 (m, 16, H-8 or Rha/Glu/Glu′-H), 2.66 (m, 2H, H-7), 0.92 (d, 3H, J = 6.1 Hz, Rha H-6). The structure was determined as echinacoside based on the results of ^{1}H NMR [38].

Target **3**: HRMS (m/z) 801.2811 $[M + H]^+$ (calcd. for C36H48O20, 801.2823). ^{1}H-NMR (DMSO-d_6, 600 MHz) δ (ppm): 7.55 (d, 1H, J = 15.6 Hz, H-7′), 7.27 (d, 1H, J = 1.7 Hz, H-2′), 7.09 (dd, 1H, J = 1.5 Hz, J = 8.5 Hz, H-6′), 6.79 (d, 1H, J = 8.0 Hz, H-5′), 6.66 (d, 1H,

J = 2.3 Hz, H-2), 6.63 (d, 1H, J = 8.2 Hz, H-5), 6.52 (dd, 1H, J = 2.2 Hz, J = 7.8 Hz, H-6), 6.39 (d, 1H, J = 15.6 Hz, H-8′), 5.03 (s, 1H, Rha H-1), 4.73 (t, 1H, J = 9.73 Hz, Glu H-4), 4.38 (d, 1H, J = 7.9 Hz, Glu′ H-1), 4.17 (d, 1H, J = 8.3 Hz, Glu H-1), 3.87 (m, 1H, H-8), 3.77 (s, 3H, Ome), 3.70–2.90 (m, 16, H-8 or Rha/Glu/Glu′-H), 2.69 (m, 2H, H-7), 0.95 (d, 3H, J = 6.1 Hz, Rha H-6). The structure was determined as Jionoside A1 based on the results of ^{1}H NMR [38].

Target **4**: HRMS (m/z) 771.2751 $[M + NH_4]^+$ (calcd. for C34H44O19, 774.2815). ^{1}H-NMR (DMSO-d_6, 600 MHz) δ (ppm): 7.44 (d, 1H, J = 16.3 Hz, H-7′), 7.01 (s, 1H, H-2′), 6.95 (d, 1H, J = 6.8 Hz, H-6′), 6.71 (d, 1H, J = 6.9 Hz, H-5′), 6.61 (d, 1H, J = 2.1 Hz, H-2), 6.62 (d, 1H, J = 8.1 Hz, H-5), 6.47 (dd, 1H, J = 1.9 Hz, J = 7.8 Hz, H-6), 6.17 (d, 1H, J = 14.2 Hz, H-8′), 5.02 (s, 1H, Rha H-1), 4.78 (d, 1H, J = 2.8 Hz, Api H-1), 4.67 (t, 1H, J = 9.6 Hz, Glu H-4), 4.35 (d, 1H, J = 8.3 Hz, Glu H-1), 3.88–3.00 (m, 15H, H-8 or Rha/Glu/Api-H), 2.65 (m, 2H, H-7), 0.96 (d, 3H, J = 6.0 Hz, Rha H-6). The structure was determined as Forsythoside B based on the results of ^{1}H NMR [38].

Target **5**: HRMS (m/z) 642.2333 $[M + NH_4]^+$ (calcd. for C29H36O15, 642.2392). ^{1}H-NMR (DMSO-d_6, 600 MHz) δ (ppm): 7.45 (d, 1H, J = 16.2 Hz, H-7′), 6.99 (d, 1H, J = 1.7 Hz, H-2′), 6.97 (dd, 1H, J = 1.7 Hz, J = 8.6 Hz, H-6′), 6.74 (d, 1H, J = 8.1 Hz, H-5′), 6.64 (d, 1H, J = 1.7 Hz, H-2), 6.62 (d, 1H, J = 8.1 Hz, H-5), 6.49 (dd, 1H, J = 1.7 Hz, J = 7.7 Hz, H-6), 6.17 (d, 1H, J = 16.2 Hz, H-8′), 5.01 (s, 1H, Rha H-1), 4.72 (t, 1H, J = 9.3 Hz, Glu H-4), 4.35 (d, 1H, J = 7.7 Hz, Glu H-1), 3.88 (m, 1H, Rha H-2), 3.78 (m, 1H, Glu H-3), 3.75 (m, 2H, H-8), 3.55 (m, 1H, Rha H-3), 3.45–3.00 (m, 6H, Rha/Glu-H), 2.65 (m, 2H, H-7), 1.01 (d, 3H, J = 5.7 Hz, Rha H-6). The structure was determined as verbascoside based on the results of ^{1}H NMR [38].

4. Conclusions

In this study, a DES was employed to fuel the extraction and counter-current separation of phenylpropanoid glycosides from *Pedicularis oederi* Vahl. After a response surface methodology optimization of ultrasonic-assisted extraction conditions, phenylpropanoid glycosides were well extracted and enriched. Under the optimal condition of 230 W ultrasonic power with a 50% water content in DESs and a liquid-to-solid ratio of 24, it yielded an extraction rate of 68.52%. Then, the extracted and enriched samples were separated by HSCCC with a DES-involved solvent system. Four phenylpropanoid glycosides, including 31.6 mg of echinacoside, 65.3 mg of Jionoside A1, 28.9 mg of Forsythoside B, 74.1 mg of verbascoside, and 21.2 mg of kaempferol-3-*O*-rutinoside were obtained from 900 mg of the sample in about 360 min. The results demonstrated DESs as alternative green solvents, widely exploitable in natural product extraction and separation.

Author Contributions: Conceptualization, D.Z. and D.L.; methodology, Y.W. (Yao Wang), J.D. and X.Z.; software, Y.W. (Yao Wang), J.D. and Y.W. (Yuhan Wang); validation, F.H.; formal analysis, Y.W. (Yao Wang), J.D. and L.L.; investigation, Y.W. (Yao Wang) and J.D.; resources, Y.W. (Yao Wang) and J.D.; data curation, D.Z.; writing—original draft preparation, Y.W. (Yao Wang), J.D., D.L. and D.Z.; writing—review and editing, Y.W. (Yao Wang), J.D., D.L. and D.Z.; visualization, Y.W. (Yao Wang) and J.D.; supervision, D.L. and D.Z.; project administration, D.L. and D.Z.; funding acquisition, D.L. and D.Z. All authors have read and agreed to the published version of the manuscript..

Funding: This research was funded by the National Natural Science Foundation of China grant number 82304892 and 32260129. And The APC was funded by 82304892.

Data Availability Statement: All data supporting the results of this study are included in the manuscript, and the datasets are available upon request.

Conflicts of Interest: The authors declare no conflicts of interest.

References

1. Tkach, N.; Ree, R.H.; Kuss, P.; Röser, M.; Hoffmann, M.H. High mountain origin, phylogenetics, evolution, and niche conservatis of arctic lineages in the hemiparasitic genus Pedicularis (Orobanchaceae). *Mol. Phylogenetics Evol.* **2014**, *76*, 75–92. [CrossRef]
2. Yatoo, M.I.; Dimri, U.; Gopalakrishnan, A.; Karthik, K.; Gopi, M.; Khandia, R.; Saminathan, M.; Saxena, A.; Alagawany, M.; Fara M.R.; et al. Beneficial health applications and medicinal values of Pedicularis plants: A review. *Biomed. Pharmacother.* **201** *95*, 1301–1313. [CrossRef]
3. Chen, C.; Zhao, X.-H.; Yue, H.-L.; Li, Y.-L.; Chen, T. Separation of Phenylpropanoid Glycosides from a Chinese Herb by HSCCC. *Chromatogr. Sci.* **2013**, *52*, 395–399. [CrossRef]
4. Jessop, P.G.; Phan, L.; Carrier, A.; Robinson, S.; Dürr, C.J.; Harjani, J.R. A solvent having switchable hydrophilicity. *Green Che* **2010**, *12*, 809–814. [CrossRef]
5. Jessop, P.G. Searching for green solvents. *Green Chem.* **2011**, *13*, 1391–1398. [CrossRef]
6. Cao, J.; Chen, L.; Li, M.; Cao, F.; Zhao, L.; Su, E. Two-phase systems developed with hydrophilic and hydrophobic deep eutect solvents for simultaneously extracting various bioactive compounds with different polarities. *Green Chem.* **2018**, *20*, 1879–188 [CrossRef]
7. Erythropel, H.C.; Zimmerman, J.B.; de Winter, T.M.; Petitjean, L.; Melnikov, F.; Lam, C.H.; Lounsbury, A.W.; Mellor, K.E.; Jankovi N.Z.; Tu, Q.; et al. The Green ChemisTREE: 20 years after taking root with the 12 principles. *Green Chem.* **2018**, *20*, 1929–196 [CrossRef]
8. Huang, J.; Guo, X.; Xu, T.; Fan, L.; Zhou, X.; Wu, S. Ionic deep eutectic solvents for the extraction and separation of natura products. *J. Chromatogr. A* **2019**, *1598*, 1–19. [CrossRef]
9. Afonso, J.; Mezzetta, A.; Marrucho, I.M.; Guazzelli, L. History repeats itself again: Will the mistakes of the past for ILs be repeate for DESs? From being considered ionic liquids to becoming their alternative: The unbalanced turn of deep eutectic solvent *Green Chem.* **2023**, *25*, 59–105. [CrossRef]
10. Wang, T.; Wang, Q.; Li, P.; Yang, H. High-Speed Countercurrent Chromatography-Based Method for Simultaneous Recovery an Separation of Natural Products from Deep Eutectic Solvent Extracts. *ACS Sustain. Chem. Eng.* **2020**, *8*, 2073–2080. [CrossRef]
11. Coskun, O. Separation techniques: Chromatography. *North. Clin. Istanb.* **2016**, *3*, 156.
12. Zou, D.; Cui, Y.; Li, S.; Sang, D.; Liu, W.; Zhao, T.; Gu, X.; Chen, T.; Li, Y. The applicability of high-speed counter-curren chromatography for preparative separation of biosynthesis products: Glycosylation products as example. *J. Sep. Sci.* **202** *44*, 4368–4375. [CrossRef]
13. Chen, T.; Jia, J.; Zou, D.; Shen, C.; Wang, S.; Gao, X.; Yun, Z.; Ma, J.; Zhao, J.; Wang, H. Nuclear magnetic resonance-based solven system selection for counter-current chromatography separation of compounds present in the same high-performance liquic chromatography peak: Flavonoids in barley seedlings as an example. *J. Sep. Sci.* **2023**, *46*, 2300406. [CrossRef]
14. Zou, D.; Du, Y.; Kuang, J.; Sun, S.; Ma, J.; Jiang, R. PH-zone-refining counter-current chromatography with a hydrophili organic/salt-containing two-phase solvent system for preparative separation of polar alkaloids from natural product *J. Chromatogr. A* **2018**, *1553*, 1–6. [CrossRef]
15. Wang, Q.; Chen, T.; Cui, Y.; Li, S.; Jiang, X.; Zhao, G.; Li, Y.; Zou, D. The applicability of pH-zone-refining counter-curren chromatography for preparative separation of biosynthesis products: Glycosylation products as example. *J. Chromatogr. A* **2021** *1657*, 462582. [CrossRef]
16. Chen, T.; Wang, Q.; Wang, Q.; La, M.; Li, Y.; He, L.; Zou, D. Ab initio calculation based solvent system selection in silico fo counter-current chromatography: Separation of resibufogenin glycosylation products. *J. Chromatogr. A* **2022**, *1686*, 463649 [CrossRef]
17. Tanimura, T.; Pisano, J.J.; Ito, Y.; Bowman, R.L. Droplet Countercurrent Chromatography. *Science* **1970**, *169*, 54–56. [CrossRef]
18. Ito, Y.; Conway, W.D. Development of countercurrent chromatography. *Anal. Chem.* **1984**, *56*, 534A–554A. [CrossRef]
19. Conway, W.D.; Ito, Y. Resolution in Countercurrent Chromatography. *J. Liq. Chromatogr.* **1985**, *8*, 2195–2207. [CrossRef]
20. Brent Friesen, J.; Pauli, G.F. GUESS—A Generally Useful Estimate of Solvent Systems for CCC. *J. Liq. Chromatogr. Rel. Technol* **2005**, *28*, 2777–2806. [CrossRef]
21. Ito, Y. Golden rules and pitfalls in selecting optimum conditions for high-speed counter-current chromatography. *J. Chromatogr. A* **2005**, *1065*, 145–168. [CrossRef]
22. Zou, D.l.; Chen, T.; Li, H.m.; Chen, C.; Zhao, J.y.; Li, Y.l. Separation of three phenolic high-molecular-weight compounds from the crude extract of Terminalia Chebula Retz. by ultrasound-assisted extraction and high-speed counter-current chromatography *J. Sep. Sci.* **2016**, *39*, 1278–1285. [CrossRef]
23. Yu, P.; Ma, J.; Du, Y.; Wang, Q.; Chen, C.; Bai, X.; He, L.; Zou, D. An efficient strategy based on liquid-liquid extraction and pH-zone-refining counter-current chromatography for selective enrichment, separation, and purification of alkaloids and organic Acids from natural products. *J. Sep. Sci.* **2020**, *43*, 3607–3614. [CrossRef]
24. Zou, D.; Zhu, X.; Zhang, F.; Du, Y.; Ma, J.; Jiang, R. An Efficient Strategy Based on Liquid–Liquid Extraction with Three-Phase Solvent System and High Speed Counter-Current Chromatography for Rapid Enrichment and Separation of Epimers of Minor Bufadienolide from Toad Meat. *J. Agric. Food Chem.* **2018**, *66*, 1008–1014. [CrossRef]
25. Li, J.; Wang, Q.; Wang, Y.; La, M.; Mian, R.; He, L.; Suonan, J.; Zou, D. An efficient strategy for large-scale preparation of low polarity gingerols directly from ginger crude extract by high-speed countercurrent chromatography with different rotation mode *J. Sep. Sci.* **2023**, *46*, 2300320. [CrossRef]

'6. Khuri, A.I.; Mukhopadhyay, S. Response surface methodology. *Wiley Interdiscip. Rev. Comput. Stat.* **2010**, *2*, 128–149. [CrossRef]
'7. Liu, G.; Feng, S.; Sui, M.; Chen, B.; Sun, P. Extraction and identification of steroidal saponins from Polygonatum cyrtonema Hua using natural deep eutectic solvent-synergistic quartz sand-assisted extraction method. *J. Sep. Sci.* **2023**, *46*, 2200823. [CrossRef]
'8. Vyviurska, O.; Koljančić, N.; Gomes, A.A.; Špánik, I. Optimization of enantiomer separation in flow-modulated comprehensive two-dimensional gas chromatography by response surface methodology coupled to artificial neural networks: Wine analysis case study. *J. Chromatogr. A* **2022**, *1675*, 463189. [CrossRef]
29. He, Q.; Lei, Q.; Huang, S.; Zhou, Y.; Liu, Y.; Zhou, S.; Peng, D.; Deng, X.; Xue, J.; Li, X. Effective extraction of bioactive alkaloids from the roots of Stephania tetrandra by deep eutectic solvents-based ultrasound-assisted extraction. *J. Chromatogr. A* **2023**, *1689*, 463746. [CrossRef]
30. Dean, A.; Voss, D.; Draguljić, D.; Dean, A.; Voss, D.; Draguljić, D. Response surface methodology. In *Design and Analysis of Experiments*; Springer: Berlin/Heidelberg, Germany, 2017; pp. 565–614.
31. Oka, F.; Oka, H.; Ito, Y. Systematic search for suitable two-phase solvent systems for high-speed counter-current chromatography. *J. Chromatogr. A* **1991**, *538*, 99–108. [CrossRef]
32. Sutherland, I.A. Recent progress on the industrial scale-up of counter-current chromatography. *J. Chromatogr. A* **2007**, *1151*, 6–13. [CrossRef]
33. Conway, W.D. Counter-current chromatography: Simple process and confusing terminology. *J. Chromatogr. A* **2011**, *1218*, 6015–6023. [CrossRef]
34. Hopmann, E.; Arlt, W.; Minceva, M. Solvent system selection in counter-current chromatography using conductor-like screening model for real solvents. *J. Chromatogr. A* **2011**, *1218*, 242–250. [CrossRef]
35. Hopmann, E.; Frey, A.; Minceva, M. A priori selection of the mobile and stationary phase in centrifugal partition chromatography and counter-current chromatography. *J. Chromatogr. A* **2012**, *1238*, 68–76. [CrossRef]
36. Hu, R.; Pan, Y. Recent trends in counter-current chromatography. *TrAC Trends Anal. Chem.* **2012**, *40*, 15–27. [CrossRef]
37. Jang, Y.S.; Wang, Z.; Lee, J.-M.; Lee, J.-Y.; Lim, S.S. Screening of Korean Natural Products for Anti-Adipogenesis Properties and Isolation of Kaempferol-3-O-rutinoside as a Potent Anti-Adipogenetic Compound from Solidago virgaurea. *Molecules* **2016**, *21*, 226. [CrossRef]
38. Zhang, L.; Yue, H.L.; Zhao, X.H.; Li, J.; Shao, Y. Separation of Four Phenylpropanoid Glycosides from a Chinese Herb by HSCCC. *J. Chromatogr. Sci.* **2014**, *53*, 860–865. [CrossRef]

separations

MDPI

Article

Development, Validation, and Greenness Assessment of Eco-Friendly Analytical Methods for the Determination of Abiraterone Acetate in Pure Form and Pharmaceutical Formulations

Erten Akbel

Faculty of Health Sciences, Usak University, Usak 64200, Turkey; erten.akbel@usak.edu.tr; Tel.: +90-5058964040

Abstract: This study presents the development and validation of environmentally friendly analytical methods for quantifying Abiraterone Acetate (AA) in both its pure form and commercial pharmaceutical formulations. An optimized High-Performance Liquid Chromatography (HPLC) method was developed using an Agilent Extend C18 column (250 mm × 4.6 mm, 5 μm) at 25 °C. The mobile phase consisted of formic acid and ethanol in isocratic mode, with a flow rate of 1.0 mL min^{-1}, and detection was performed at 253 nm. The spectrophotometric method involved a comprehensive evaluation of AA's spectral properties in various solvents, with ultrapure water providing the most suitable spectra for analysis at 253 nm. Both methods were validated according to ICH guidelines, demonstrating selectivity, linearity, accuracy, precision, detection and quantification limits, and robustness, with correlation coefficients exceeding 0.999 across the 5–30 μg mL^{-1} concentration range. Comparative statistical analysis using Student's *t*-test and Fisher's F-test showed no significant differences between the two methods. The environmental impact of both methods was assessed using AGREE and GAPI software, confirming their sustainability. These validated methods offer reliable and eco-friendly approaches for the quantitative analysis of AA in tablet formulations, promoting safer and greener laboratory practices in pharmaceutical analysis.

Keywords: Abiraterone Acetate; green analytical methods; HPLC; spectrophotometry; pharmaceutical validation; sustainable laboratory practices

1. Introduction

In the face of today's growing environmental concerns, the shift towards more sustainable analytical methodologies is paramount. This trend holds particular relevance for liquid chromatography techniques, which are widely used but often lack alignment with green chemistry principles. Specifically, principles 1, 5, 6, 11, and 12 of green chemistry are pertinent to analytical methods. These principles advocate for safer chemistry to minimize waste, energy efficiency, real-time analysis to prevent pollution, and accident prevention [1]. When developing a new analytical method for detecting a specific analyte, two crucial factors must be addressed. First, the method's metrological value must be validated to ensure accuracy, precision, and reproducibility. Second, environmental sustainability must be prioritized by incorporating safer, less harmful chemicals and processes, and adhere to the guidelines for green chemistry. The application of these principles is critical yet still underutilized in many fields, including pharmaceutical analysis [2].

High-Performance Liquid Chromatography (HPLC) remains a cornerstone technology in pharmaceutical quality control, owing to its robust and reliable performance. Traditionally, reverse-phase HPLC utilizes a polar mobile phase combined with a hydrophobic stationary phase. The mobile phase often consists of a mixture of water, modified with additives, and organic solvents like acetonitrile or methanol. In reverse phase separation mode, mobile phases primarily composed of water and aqueous buffers may be utilized, depending on the analyte's solubility. However, using only water and aqueous buffers in the mobile phase can lead to prolonged analysis times. Consequently, mobile phases that combine aqueous buffers with organic modifiers are preferred for more efficient separation. These organic solvents are preferred due to their favorable chromatographic characteristics, such as miscibility with water, low viscosity, and chemical stability, which ensure high-quality analytical performance [3]. Despite these advantages, both acetonitrile and methanol pose significant environmental and health risks. Acetonitrile is highly volatile, flammable, and toxic, while methanol, though less toxic and biodegradable, is still classified as hazardous [4]. The widespread use of these solvents in HPLC has resulted in substantial amounts of waste, leading to significant environmental and disposal challenges [5].

As HPLC usage expands, so does the generation of solvent waste, amplifying the need for more sustainable approaches. This necessity has pushed researchers toward the development of greener analytical methods, focusing on the reduction or elimination of toxic chemicals in favor of safer, environmentally benign alternatives. One such alternative is ethanol, which has emerged as a promising candidate for use in green HPLC methods. Ethanol's lower toxicity, biodegradability, and cost-effectiveness make it an attractive option, particularly for laboratories with limited resources or those in developing regions [6]. Moreover, ethanol offers comparable chromatographic performance to acetonitrile and methanol, exhibiting similar selectivity and retention times, making it a viable green alternative without compromising analytical efficiency [7]. Additionally, the eluent strengths for a C18 column are comparable: acetonitrile and ethanol both have strengths of 3.1, while methanol has a strength of 1.0 [8]. These properties further substantiate ethanol's potential as a substitute for methanol and acetonitrile in HPLC applications.

However, the implementation of ethanol in HPLC is not without its challenges. Ethanol has a higher ultraviolet (UV) cutoff value than acetonitrile and methanol, which can lead to increased background noise and reduced sensitivity in UV detection, particularly in gradient elution systems [9]. Additionally, ethanol's higher viscosity compared to acetonitrile and methanol can result in elevated column pressures, potentially affecting both the efficiency and lifespan of chromatographic columns. These issues can be mitigated by optimizing the HPLC method, such as by adjusting column temperature to reduce ethanol's viscosity, thus maintaining system performance and longevity [10].

In this study, I aim to address the growing need for environmentally sustainable analytical techniques by developing a green liquid chromatography (Green LC) method utilizing ethanol as the mobile phase solvent. Ethanol's favorable environmental profile and chromatographic performance make it an ideal candidate for promoting greener analytical practices [11]. Furthermore, I introduce a Simplified Spectrophotometric Method for the analysis of AA, an anti-androgen drug used primarily for treating prostate cancer [12]. AA is also being explored for its potential use in treating ovarian and breast cancers [13]. Its chemical properties and growing therapeutic applications highlight the need for reliable, sustainable analytical methods for its quantification in pharmaceutical formulations [14]. The chemical features of AA are presented in Table 1, as already detailed elsewhere [15].

The literature reveals a wide range of analytical methods for quantifying AA in pharmaceutical and biological samples, including spectrofluorimetric, spectrophotometric, and liquid chromatography techniques [16]. Many different analytical methods have been developed for the quantification of AA in biological fluids, bulk, and dosage forms. For example, the Liquid chromatographic-tandem mass spectroscopy method for the quantification of AA in human plasma [17]; the RP-HPLC/UV technique for the quantification of AA in rat plasma and its application to a pharmacokinetic study in rats [18]; the LC-MS/MS-ESI method for the determination of AA in rat and human plasma and its application to a pharmacokinetic study [19]; Liquid chromatography coupled with ultraviolet detection and electron ionization mass spectrometry method for the analysis of AA stress degradation behavior [20]; AA in human plasma was determined using an ultra-high-performance liquid chromatography-tandem mass spectrometry technique [21]; and AA in human plasma was determined using an HPLC-fluorescence technique [22]. While effective, these methods often require expensive equipment, toxic solvents, and complex procedures, limiting their accessibility and increasing their environmental impact [15]. The introduction of greener methods for AA analyses is, therefore, timely and necessary. In this context, our study presents a novel, environmentally friendly Green LC method using ethanol, as well as a Simplified Spectrophotometric Method that employs ultrapure water as the solvent [23]. These methods aim to reduce the environmental burden associated with traditional HPLC techniques while maintaining robust analytical performance [24].

Table 1. The physicochemical properties of AA.

Property	Value
Name	17-(pyridin-3-yl)androsta-5,16-dien-3β-yl acetate
Formula	$C_{26}H_{33}NO_2$
Structure	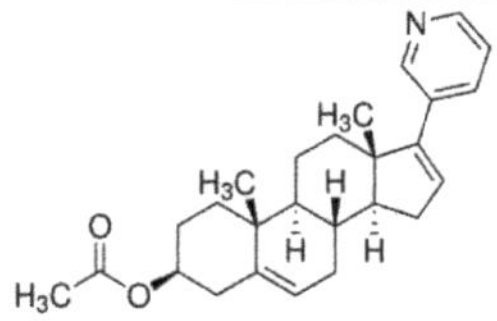
Molecular weight	391.55 Daltons
Melting point	144–145 °C
Log P	3.97
pKa/Strongest acidic	18.32
pKa/Strongest basic	4.81
Solubility	Soluble in organic solvents such as ethanol (28 mg/mL at 25 °C), dimethylformamide (~16 mg/mL), and dimethylsulfoxide (~2 mg/mL). Very slightly soluble in aqueous buffers (<1 mg/mL at 25 °C).

2. Materials and Methods

2.1. Instruments

Chromatographic analyses were performed using an Agilent 1260 system (Agilent Technologies, Palo Alto, CA, USA), which includes a UV-Vis Detector, a Quaternary Pump, a Vacuum Degasser, a Column Oven, and ChemStation Software (B.04.03 (November, 2010)) for data acquisition and analysis. The chromatographic separation was achieved with an Agilent Extend C18 column (250 mm × 4.6 mm, 5 μm). Spectrophotometric analyses were conducted with a UV-1800 dual-beam spectrophotometer (Shimadzu, Kyoto, Japan), using 1.0 cm quartz cuvettes and UV-Probe Software (version 2.70) for data analysis. pH measurements were made with a Mettler-Toledo pH meter (Mettler Toledo, Greisensee, Switzerland) equipped with a glass electrode. Ultrapure water for the experiments was produced using a Millipore Milli-Q system (Millipore, Bedford, MA, USA).

2.2. Reagents and Chemicals and Materials

All solvents utilized were of gradient-grade purity suitable for liquid chromatography, including acetonitrile (≥99.9%), methanol (≥99.0%), ethanol (≥99%), and formic acid (≥99.0%). These solvents were sourced from Sigma-Aldrich Chemie GmbH (Istanbul, Turkey). The United States Pharmacopeia (USP) reference standard of Abiraterone Acetate (AA) used in the study had a purity of ≥99.5%. AA tablets (Zytiga, 250 mg) were obtained from a local pharmacy in Afyonkarahisar, Turkey. Ultrapure water, with a resistivity of 0.080 μS cm^{-1}, was used for the preparation of all solutions and the mobile phase. The mobile phase was filtered through a 0.20 μm membrane filter using a vacuum pump and was sonicated prior to the analyses to ensure optimal purity and performance during chromatography.

2.3. Pharmaceutical Solutions

The Stock Standard Solution (500 µg mL^{-1}) was prepared by accurately weighing 25 mg of the USP reference standard AA and transferring it to a 50 mL volumetric flask. Ultrapure water (20 mL) was added, and the mixture was sonicated for 6 min to achieve complete dissolution. The volume was then adjusted to 50 mL with ultrapure water. Standard solutions (5, 10, 15, 20, 25, and 30 µg mL^{-1}) were prepared by performing serial dilutions of the stock standard solution with ultrapure water to achieve the desired concentrations within the range of 5 to 30 µg mL^{-1}. The sample solution (25 µg mL^{-1}) was prepared by accurately weighing ten Zytiga tablets (250 mg each) and recording the average mass per tablet. The tablets were crushed in a dry, clean mortar and ground to a fine powder, which was thoroughly mixed. A 50 mg portion of the powdered tablet, equivalent to AA, was transferred to a 100 mL volumetric flask. Approximately 40 mL of ultrapure water was added, and the mixture was shaken on a rotary shaker for 30 min to ensure complete dissolution. The volume was adjusted to 100 mL with ultrapure water, and the solution was sonicated for 10 min. Finally, the solution was filtered through a 0.45 µm membrane filter to obtain the stock sample solution, which was diluted with ultrapure water to achieve a final concentration of 25 µg mL^{-1}.

2.4. Determination of λ_{Max}

Six standard solutions (5, 10, 15, 20, 25, 25, 30 µg mL^{-1}) were prepared from the stock standard solution in the concentration range of 5–30 µg mL^{-1}, each with three parallels. These solutions were analyzed using a UV-Vis spectrophotometer (Shimadzu, Kyoto, Japan) in the wavelength range of 200 to 400 nm. The spectrophotometer was calibrated using a blank solution (solvent only) before each scan. The maximum absorbance (λ_{Max}) for AA was determined by analyzing the absorbance spectra and identifying the wavelength at which the highest absorbance occurred. All measurements were conducted at room temperature, and each concentration was scanned in triplicate to ensure accuracy and reproducibility.

2.5. Development of Methods and Optimization of Conditions

Chromatographic conditions were carefully optimized to achieve superior peak parameters, including well-defined peak shapes, minimal tailing factors, reduced retention times, and high theoretical plate numbers. Initial experiments involved the evaluation of various mobile phases with different buffer systems, but none met the necessary system compatibility requirements. Additionally, several types and lengths of columns were tested, but these also failed to provide the desired system compatibility parameters.

Optimal peak parameters were obtained using an Extend C18 column (250 mm × 4.6 mm, 5 µm). Several mobile phase compositions were evaluated, including mixtures of water/methanol, water/acetonitrile, and water/ethanol. Initially, a mobile phase composed of acetonitrile and ultrapure water (20/80, *v*/*v*) resulted in excessively prolonged analysis times. To shorten the analysis time, the water component was acidified with formic acid to a pH of 2.0. This pH was selected because it is significantly below the pKa of the analyte, ensuring that the analyte remains protonated and soluble, which is crucial for achieving optimal retention and separation on the C18 column. Under these optimized conditions, the sample solution was injected to assess potential impurities that could interfere with the analyte peak or residual drug matrix components that might persist on the column. Sequential sample injections were performed, with a total analysis time of 6 min. No carryover of impurities was observed between injections, confirming the suitability of the 6 min analysis time. The column temperature was maintained at 25 °C, selected for its ability to enhance column efficiency, maintain low column pressure, improve peak shape, and ensure cost-effectiveness. Spectral characterization was performed during the calibration process, focusing on critical factors that influence the accuracy of spectrophotometric methods, including wavelength accuracy, spectral bandwidth, stray light, and linearity. These characteristics are essential for achieving reliable and reproducible analytical results.

For spectrophotometric analysis, the spectral characteristics of AA were assessed in various solvents, including ultrapure water, ethanol, methanol, and isopropyl alcohol. Ultrapure water was selected as the solvent as it provided the most favorable spectral properties for AA. Standard solutions of AA exhibited maximum absorbance at 253 nm, and absorbance measurements for both the standard and sample solutions were recorded at this wavelength.

2.6. Validation of Analytical Methods

The analytical methods developed for the quantification of AA were rigorously validated according to the ICH Q2 (R1) guidelines [6,25]. The validation process included the following parameters: (1) Selectivity was assessed to ensure the method's ability to accurately identify and quantify AA in the presence of other substances. (2) System suitability was verified to confirm that the chromatographic system functioned properly and met predefined criteria. (3) Linearity was evaluated to determine the method's capability to produce results proportional to the concentration of AA within the specified range. (4) Precision was analyzed by examining both repeatability and reproducibility under consistent conditions. (5) Sensitivity was measured by identifying the lowest concentration of AA that could be reliably detected and quantified. (6) Robustness was tested to evaluate the method's reliability under minor variations in operational conditions.

2.7. Selectivity and System Suitability of Methods

To evaluate the selectivity of the chromatographic method, solutions of standard, sample, and mobile phase were injected into the HPLC system. The retention time (Rt) of AA in the chromatograms of both commercial formulations and sample solutions was compared to that of the standard solution to ensure accurate identification and quantification of AA. The system suitability of the chromatographic method was assessed by injecting a 25 μg mL^{-1} standard solution of AA into the system at six regular intervals. The following parameters were recorded from the chromatograms: peak area, retention time, tailing factor, and number of theoretical plates. The relative standard deviation (RSD) values for peak area and retention time were calculated to evaluate the precision and consistency of the method.

The selectivity of the spectrophotometric method was assessed by scanning the spectra of standard solutions, sample solutions, and the solvent (ultrapure water) within the wavelength range of 200 to 400 nm. The spectra obtained were analyzed to identify any potential interfering bands and to confirm the absence of overlaps or interferences. For evaluating the system's suitability, absorbance values of a 25 μg mL^{-1} standard solution of AA were recorded in six separate measurements. The RSD of these absorbance values was computed to determine the precision and reliability of the method.

2.8. Evaluation of Linearity of Chromatographic and Spectrophotometric Methods

The linearity of the chromatographic method was evaluated by injecting six standard solutions with concentrations ranging from 5 to 30 μg mL^{-1} into the HPLC system. This procedure was conducted in triplicate over three separate days. Peak areas corresponding to each concentration were recorded, and calibration curves were constructed by plotting concentrations on the x-axis against peak areas on the y-axis. Regression analysis was performed using the least squares method. Linearity was assessed based on the absolute mean recovery, RSD, and the R-value of the calibration curve.

Similarly, the linearity of the spectrophotometric method was determined by measuring the absorbance values of six standard solutions with concentrations ranging from 5 to 30 μg mL^{-1}. These measurements were taken in triplicate across three different days, and absorbance values were recorded for each concentration. Calibration curves were plotted with concentrations on the x-axis and absorbance values on the y-axis. Regression analysis was carried out using the least squares method. Linearity was evaluated by analyzing the absolute mean recovery, RSD, and R-value of the calibration curve.

2.9. Evaluation of Sensitivity

The sensitivity of the chromatographic and spectrophotometric methods was evaluated by calculating the limits of detection (LOD) and limits of quantification (LOQ). These limits were established using the standard deviations of the intercept and slope of the calibration curves, following the methodologies outlined by according to Zhao et al. [26] and Shetty and Shah [27]. The calculations were performed using the following equations:

$$\text{LOD} = (3.3 \times \sigma_{\text{intercept}})/\text{Slope}$$

where $\sigma_{\text{intercept}}$ is the standard deviation of the intercept of the calibration curve.

The LOD and LOQ were calculated for both methods to assess their ability to detect and quantify low concentrations of AA reliably.

2.10. Precision

The precision of the analytical methods was evaluated by examining both intra-day and inter-day variability. A standard solution of AA at a concentration of 20 µg mL^{-1} was analyzed multiple times within a single day. Three replicates (n = 3) were conducted for each method (chromatographic and spectrophotometric). Peak areas and retention times were recorded, and RSD values for these parameters were computed. Absorbance values were also measured, and their RSD values were calculated.

Additionally, a standard solution of AA at a concentration of 25 µg mL^{-1} was analyzed on three separate days. In total, nine replicates (n = 9) were performed for each method. Peak areas and retention times were recorded, and RSD values for these measurements were determined. Absorbance values were measured, and their RSD values were calculated. This evaluation confirmed that the methods yielded consistent and reproducible results both within a single day and across multiple days.

2.11. Accuracy

The accuracy of the analytical methods was evaluated using the standard addition method to determine the percentage recovery of AA. The procedure included the following steps: Preparation of Solutions, where a pre-analyzed AA solution at a concentration of 10 µg mL^{-1} was utilized. Additional AA standards were then introduced to achieve final concentrations of 25 µg mL^{-1}, corresponding to 75%, 100%, and 125% of the analyte added. Analysis was conducted by re-quantitatively analyzing the spiked solutions using the developed chromatographic and spectrophotometric methods. The amount of AA recovered from each spiked solution was determined, and the percentage accuracy was calculated using the following formula:

$$\text{Percent Recovery} = \left(\frac{\text{Amount Recovered} - \text{Amount Initially Present}}{\text{Amount added}} \right)$$

The process was repeated six times (n = 6) for each concentration level to ensure reliability.

2.12. Robustness

The robustness of the chromatographic method was evaluated by introducing minor, deliberate modifications to the method conditions. Adjustments were made to the mobile phase flow rate (±0.1 mL min^{-1}), variations in the organic modifier content of the mobile phase were tested (±2%), and the detection wavelength was altered (±2 nm). Following each modification, a standard solution of AA (25 µg mL^{-1}) was injected into the chromatographic system. System suitability parameters, including peak area, retention time, tailing factor, and number of theoretical plates, were recorded and compared with results obtained under the original conditions. Each modification was tested with three replicate analyses of the standard solution to ensure consistency.

The robustness of the spectrophotometric method was assessed by implementing minor changes, including substituting ethanol with isopropyl alcohol and adjusting the detection wavelength (251 and 253 nm). Absorbance values of a standard solution (25 μg mL^{-1}) were measured under each modified condition, and results were compared with those obtained under the original spectrophotometric conditions. Each modification was tested with three replicate analyses of the standard solution to ensure accuracy. This evaluation confirmed that the methods maintained their performance despite minor changes in experimental conditions.

2.13. Application of Analytical Methods to Commercial Formulations

To validate the applicability of the proposed HPLC and spectrophotometric methods for real-world samples, commercial formulations of AA were analyzed. Specifically, the tablet form of AA (Zytiga, 250 mg) was examined using both methods to assess their performance in a practical setting. The chromatographic method involved the analysis of the commercial tablets using the developed HPLC technique. Similarly, the spectrophotometric method involved the analysis of the same tablets using UV spectrophotometry. The parameters evaluated included assay, purity, and content uniformity of the formulations.

2.14. Greenness Evaluation of Analytical Methods

Various techniques are being developed to assess the sustainability of newly established analytical methods in line with the recent advancements in green analytical chemistry (GAC) principles. Among these tools, the Analytical GREEnness (AGREE) metric and the Green Analytical Procedure Index (GAPI (2018)) have gained prominence. Both metrics use a color-coded system (red, yellow, and green) to evaluate and communicate the environmental impact of different analytical approaches [28–30].

2.15. Statistical Analysis

The results from both analytical methods were statistically compared to evaluate their consistency, accuracy, and reliability. A Student's *t*-test was employed to compare the mean values obtained from each method, assessing any significant differences between them. Additionally, the variability of the results was evaluated by comparing their standard deviations, providing insights into the precision of each method. This statistical evaluation demonstrated the effectiveness and robustness of the proposed methods for quantifying AA in commercial pharmaceutical formulations.

3. Results

3.1. Determination of the Wavelength (λ_{Max})

The overlaid UV-Vis absorption spectra AA standard solutions in the concentration range of 5–30 μg mL^{-1} are presented in Figure 1. The spectra exhibited a clear absorption maximum (λ max) at 253 nm for all concentrations. This peak intensity increased proportionally with higher concentrations of AA, confirming the detection wavelength at 253 nm. This wavelength was identified as the most suitable for further analyses due to its consistent and distinct absorbance across all tested concentrations.

3.2. Development of Methods

The absorbance spectra of AA in different solvents (ultrapure water, ethanol, methanol, and isopropyl alcohol) are illustrated in Figure 2. Ultrapure water displayed the highest clarity with the most distinct spectral profile, maintaining strong absorbance with minimal background noise. In contrast, ethanol exhibited a lower absorbance and significant baseline interference. Methanol and isopropyl alcohol also produced clear spectra but with slightly reduced absorbance compared to ultrapure water. Based on these observations, ultrapure water was selected as the optimal solvent for subsequent UV spectrophotometric analysis due to its superior absorbance and minimal interference across the tested wavelengths.

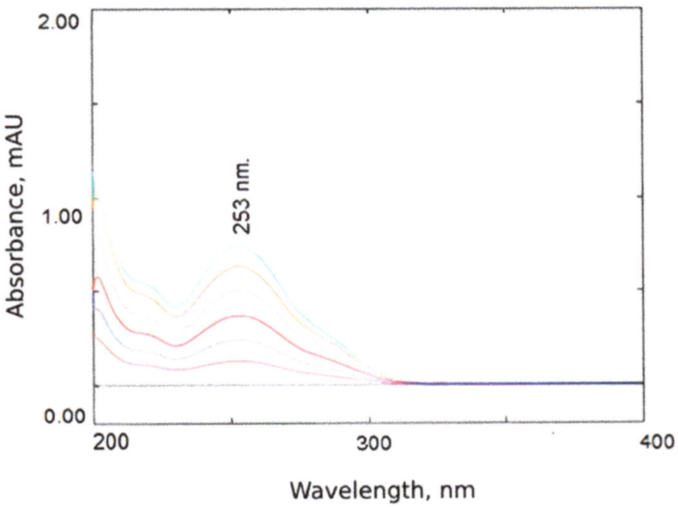

Figure 1. The overlaid spectrum of AA standard solutions in the concentration range of 5–30 μg mL^{-1}.

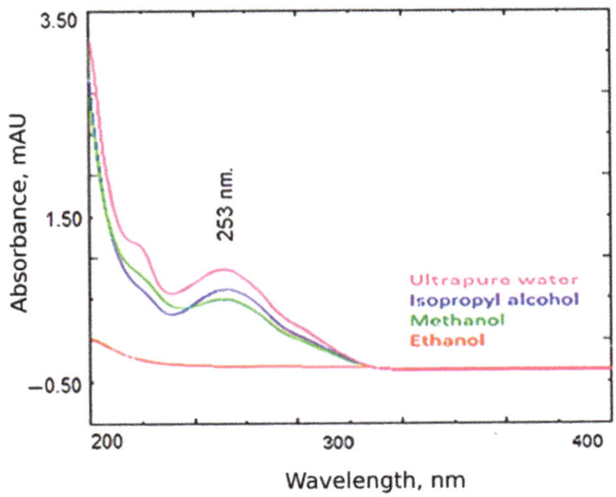

Figure 2. The overlaid spectrum obtained using different solvents (ultra-pure water, ethanol, methanol, and isopropyl alcohol).

3.3. Selectivity Evaluation for the Chromatographic Method

The chromatograms of the standard solution (30 μg mL^{-1}), sample solution (25 μg mL^{-1}), placebo, and ultra-pure water are shown in Figure 3. The standard and sample solutions exhibited a sharp and well-resolved peak at approximately 4.0 min retention time. The standard solution displayed a higher peak intensity (approximately 80 mAU) compared to the sample solution (~60 mAU), indicating a higher concentration of the analyte. In contrast, the placebo and ultra-pure water chromatograms showed no peaks, confirming the specificity of the method and the absence of interference from excipients or solvents. The peak symmetry and retention time consistency between the standard and sample solutions demonstrate the reliability and reproducibility of the analytical method employed.

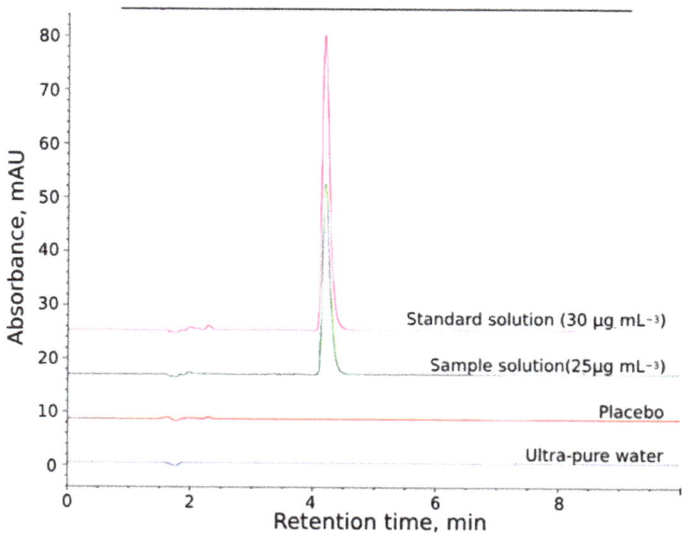

Figure 3. Chromatogram of standard, sample, placebo, and ultra-pure Water.

3.4. *Selectivity Evaluation for the Spectrophotometric Method*

Figure 4 presents the UV absorption spectra of three solutions: a standard solution (30 μg mL^{-1}), a sample solution (25 μg mL^{-1}), and ultra-pure water, measured over a wavelength range of 200 to 400 nm. Both the standard and sample solutions display similar spectral profiles, characterized by two distinct absorption peaks. The first and most pronounced peak occurs within the range of 215–220 nm, followed by a broader and less intense peak between 260–270 nm. The standard solution, with its higher concentration, exhibits marginally greater absorbance values than the sample solution throughout the spectrum. Below 210 nm, both the standard and sample spectra show a sharp increase in absorbance, indicative of the presence of the analyte. In contrast, the spectrum for ultra-pure water remains consistently flat, with absorbance values close to zero across the entire wavelength range, effectively serving as a reference baseline.

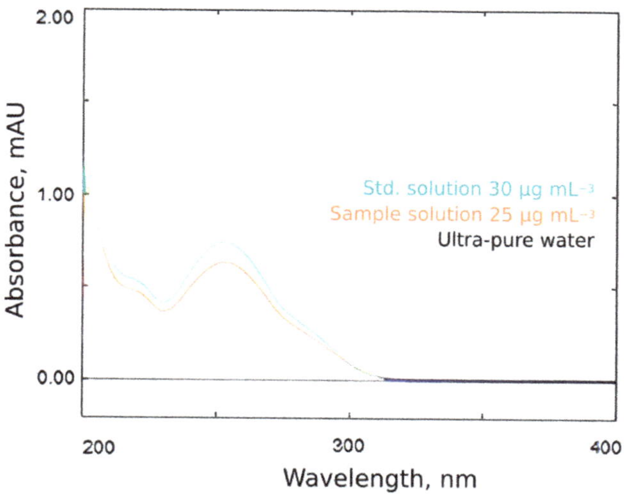

Figure 4. Spectrum of standard, sample, and ultra-pure water.

3.5. System Suitability

Table 2 shows the results of the system suitability test for both liquid chromatography and UV spectrophotometry techniques. The average retention time for the liquid chromatography was 4.197 min, with minimal deviation (RSD = 0.211%), ensuring consistent chromatographic performance. The peak area had an average of 341.02, and the theoretical plate count averaged 5671, indicating high column efficiency. UV spectrophotometry showed an average absorbance of 0.626 with an RSD of 1.094%, showing good consistency and precision.

Table 2. The results of the system suitability test.

Sample	LCT				UVST
	PA	RT (min)	PT	TPC	Absorbance
1	340.08	4.204	1.542	5666	0.632
2	339.72	4.194	1.558	5657	0.619
3	339.85	4.207	1.545	5674	0.620
4	341.61	4.186	1.548	5683	0.631
5	342.84	4.189	1.548	5656	0.621
6	342.04	4.204	1.561	5687	0.634
Average value	341.02	4.197	1.550	5671	0.626
SD	1.315	0.009	0.008	13.065	0.007
RSD	0.386	0.211	0.484	0.230	1.094

LCT, liquid chromatography technique; UVST, UV spectrophotometry technique; PA. peak area; RT, retention time; PT, peak tailing; TPC, tropic plate count; SD, standard deviation; RSD, relative standard deviation.

3.6. Linearity and Sensitivy

Table 3 summarizes the regression data for both analytical methods. The liquid chromatography technique had a linearity range of 5–30 μg/mL with a slope of 13.793 and a coefficient of determination (R^2) of 0.9999. The UV spectrophotometry technique also displayed strong linearity ($R^2 = 0.9998$) with a slightly lower slope of 0.0252. The lower LOD and LOQ were better for the liquid chromatography method (LOD = 0.40 μg/mL, LOQ = 1.10 μg/mL) compared to UV spectrophotometry (LOD = 0.70 μg/mL, LOQ = 2.00 μg/mL).

Table 3. Regression data of analytical methods.

Parameter	LCT	UVST
Linearity range/μg mL^{-1}	5–30	5–30
Regression equation ($y = mx + n$)		
Slope (m)	13.793	0.0252
Intercept (n)	−3.187	−0.0022
Coefficient of determination (r^2)	0.9999	0.9998
LOD/μg mL^{-1}	0.40	0.70
LOQ/μg mL^{-1}	1.10	2.00
Recovery % [$n = 6$]	99.59–100.58	98.57–100.48

For both analytical methods, a calibration graph was constructed by plotting absorbance values (for the spectrophotometric method) and peak areas (for the chromatographic method) against the concentration of the standard solution. The linearity of each method was evaluated using regression analysis; LCT, liquid chromatography technique; UVST, UV spectrophotometry technique; LOD, limits of detection; LOQ, limits of qualification.

3.7. Precision

Table 4 provides precision results for both methods. The intraday precision for liquid chromatography had an average retention time of 4.201 min and a peak area of 341.15, with RSDs of 0.101% and 0.225%, respectively. Similarly, UV spectrophotometry displayed an intraday absorbance RSD of 0.236%. Inter-day precision for both methods was also high, indicating robustness and repeatability.

Table 4. Precision results of analytical methods.

Precision		LCT			UVST	
		RT (min)	PA	Assay (%)	Absorbance	Assay (%)
Intraday	Average value	4.201	341.15	100.04	0.624	100.11
	SD	0.004	0.766	0.225	0.001	0.236
	RSD	0.101	0.225	0.225	0.236	0.236
Inter day	Average value	4.203	341.08	100.12	0.933	100.24
	SD	0.006	1.199	0.352	0.002	0.327

UVST, UV spectrophotometry technique; PA. peak area; RT, LCT, liquid chromatography technique; retention time; SD, standard deviation; RSD, relative standard deviation.

3.8. Accuracy

Table 5 highlights the accuracy results. For liquid chromatography, the average recovery rates across different standard addition levels (75%, 100%, 125%) ranged from 99.49% to 99.86%, with low RSD values (<0.3%). The UV spectrophotometry method also demonstrated good recovery rates, although the RSD values were slightly higher than those of liquid chromatography, especially at the 75% addition level (RSD = 0.476%).

Table 5. Accuracy results of analytical methods.

Technique	Standard Addition Level (%)	Standard Addition Amount ($\mu g\ mL^{-1}$)	Average Recovery (%)	SD	RSD
LCT	75	15	99.49	0.281	0.282
	100	20	99.72	0.189	0.190
	125	25	99.86	0.131	0.131
UVST	75	15	99.31	0.473	0.476

LCT, liquid chromatography technique; UVST, UV spectrophotometry technique; SD, standard deviation; RSD, relative standard deviation.

3.9. Robustness Evaluation

Table 6 outlines the robustness of both analytical methods under various system conditions. For liquid chromatography, recovery percentages remained within 99.44% to 100.01% across different flow rates, ethanol content, and detection wavelengths. Similarly, the UV spectrophotometry method maintained high recovery percentages (99.09%–99.82%) across different solvents and detection wavelengths, demonstrating both methods' robustness.

Table 6. The results of robustness tests for analytical methods (n = 3).

Method	System Conditions	Values	Average Recovery (%)	RSD
LCT	Normal conditions		100.01	0.47
	The high flow rate (mobile phase)	1.10 mL min^{-1}	99.59	0.54
	The low flow rate (mobile phase)	0.90 mL min^{-1}	99.44	0.58
	High detection wavelength	255 nm.	99.62	0.40
	Low detection wavelength	251 nm.	99.67	0.50
	The high ethanol content (mobile phase)	62%	99.64	0.43
	The low ethanol content (mobile phase)	58%	99.51	0.46
UVST	Normal conditions		99.82	0.64
	High detection wavelength	255 nm.	99.45	0.62
	Low detection wavelength	251 nm.	99.26	0.73
	Solvent	Ethanol	99.37	0.68
	Solvent	Isopropyl alcohol	99.09	0.66

LCT, liquid chromatography technique; UVST, UV spectrophotometry technique; RSD, relative standard deviation.

3.10. Application of Analytical Methods to Pharmaceutical Formulations

Table 7 provides a statistical evaluation of AA tablet analysis using both methods. Both techniques yielded consistent assay results, with average values near 100% for both methods (liquid chromatography: 100.00%, UV spectrophotometry: 100.00%). The t-test and F-test values indicated no significant differences between the two methods.

Table 7. Statistical evaluation of analysis results of AA tablets (Zytiga, 250 mg).

Sample	LCT		UVST	
	mg in Tablet	Assay (%)	mg in Tablet	Assay (%)
1	249.770	99.93	250.27	100.10
2	249.620	99.87	249.95	99.97
3	250.030	100.04	250.17	100.06
4	249.990	100.02	249.98	99.98
5	249.930	100.00	250.03	100.00
6	250.310	100.15	249.74	99.89
Average value	249.94	100.00	250.02	100.00
SD	0.236	0.094	0.184	0.074
RSD	0.094	0.094	0.074	0.074
t_{value}/t_{table}	2.51/6.39			
F_{value}/F_{table}	0.013/2.78			

To evaluate the consistency and reliability of the two methods, statistical tests were performed: Student's *t*-test Used to compare the mean results of the chromatographic and spectrophotometric methods. The calculated t-values were compared with tabulated values at a significance level of 0.05. The results indicated that the means of both methods were not significantly different from each other. Fisher's F-Test Applied to assess the variability in results (standard deviations) between the two methods. The calculated F-values were compared with tabulated values to determine if there were significant differences in precision. LCT, liquid chromatography technique; UVST, UV spectrophotometry technique; SD, standard deviation; RSD, relative standard deviation.

3.11. Greenness Evaluation of Analytical Methods

Figure 5 illustrates the AGREE pictograms comparing the environmental impact of three analytical methods: the USP official method (A), the chromatographic method (B), and the spectrophotometric method (C). The AGREE score for the USP official method (0.58) indicates a lower adherence to green chemistry principles, primarily due to the use of hazardous solvents and higher waste generation. In contrast, the chromatographic method, with an AGREE score of 0.78, demonstrates significant improvement in environmental performance, largely due to the substitution of ethanol as a more sustainable solvent. The spectrophotometric method achieves the highest AGREE score (0.81), reflecting its use of ultrapure water as the solvent, which minimizes waste and exposure to harmful chemicals.

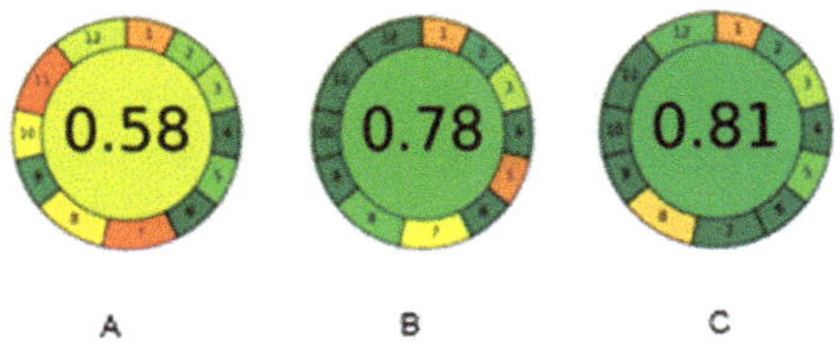

Figure 5. AGREE pictograms comparing the environmental impact of three analytical methods. (**A**) AGREE pictogram of the USP official method, (**B**) AGREE pictogram of the chromatographic method, (**C**) AGREE pictogram of the spectrophotometric method. In (**C**), Operation 3 is represented by the green color, reflecting the use of ultrapure water as an environmentally friendly solvent.

Figure 6 shows a comparison of two green analytical chemistry pictograms, labeled A (88) and B (89), which appear to represent GAPI assessments for two different analytical methods—a chromatographic method and a spectrophotometric method for AA determination. Both pictograms feature a similar structure with pentagon shapes arranged around a central red hexagon, enclosed by a green circular outline. The key difference between the two is that pictogram A contains an additional yellow pentagon segment compared to pictogram B. The numerical values (88 and 89) likely represent greenness scores for each method, with the spectrophotometric method (B) scoring slightly higher, suggesting it may be marginally more environmentally friendly or sustainable according to GAPI criteria. These visualizations effectively allow for quick comparison of the relative "greenness" of the two analytical procedures.

Figure 6. (**A**) GAPI pictogram of the chromatographic method, (**B**) GABI pictogram of the spectrophotometric method.

4. Discussion

The findings of this study demonstrate the efficacy and environmental benefits of the proposed Green LC and Simplified Spectrophotometric Methods for the quantification of AA in pharmaceutical formulations. These methods were rigorously validated, showing strong consistency with established techniques such as reverse-phase HPLC and LC-MS/MS, which are commonly used for AA analysis [23,31]. However, what sets this study apart is the integration of green chemistry principles, which prioritize sustainability alongside analytical performance.

The use of ethanol as the organic solvent in the Green LC method proved highly effective, with chromatographic performance comparable to that of traditional solvents like acetonitrile and methanol [3]. This is supported by previous research that highlights ethanol's acceptable eluotropic strength and reduced environmental footprint, making it a suitable alternative for HPLC applications [9]. Despite ethanol's slightly higher viscosity, the method remained efficient, with column pressure easily managed through minor temperature adjustments [10]. Furthermore, the retention times were notably shorter compared to conventional methods, a significant improvement in reducing solvent consumption and waste generation. These findings underscore ethanol's potential as a sustainable solvent without sacrificing analytical quality. These results underscore the superior environmental sustainability of the chromatographic and spectrophotometric methods compared to the conventional USP approach, positioning them as viable green alternatives for pharmaceutical analysis [32].

The Simplified Spectrophotometric Method, employing ultrapure water as the solvent, provided excellent sensitivity for AA quantification despite ethanol's higher UV cutoff relative to acetonitrile and methanol. The use of ultrapure water eliminates the need for hazardous organic solvents, addressing both environmental and safety concerns. This eco-friendly approach aligns with the principles of green analytical chemistry, emphasizing reduced environmental impact without compromising method sensitivity. By utilizing methods that minimize the use of harmful solvents, we enhance the sustainability of our analytical practices. Additionally, the ability to achieve reliable quantification of AA at lower wavelengths adds further merit to this green analytical approach [7,24].

One of the most significant advantages of these methods is the reduction in hazardous waste. By excluding acetonitrile and methanol, the developed techniques not only lower disposal costs but also reduce the environmental hazards associated with conventional solvents [9]. This makes the methods particularly attractive for resource-limited laboratories, where access to high-purity organic solvents can be restricted. Furthermore, ethanol's low cost and availability, especially in developing countries, further support its widespread adoption [9].

The statistical comparison using Student's t-test confirmed the consistency and reliability of the proposed methods. Both the Green Liquid Chromatography and Simplified Spectrophotometric Methods produced results comparable to those obtained using traditional methods, with no statistically significant differences in the mean values [23]. The standard deviations were also similar, indicating that the variability of the results was well within acceptable limits, further validating the precision of the methods [31]. These findings reinforce the feasibility of integrating greener solvents into routine analytical procedures without compromising accuracy or reproducibility.

5. Conclusions

In conclusion, the methods developed in this study provide sustainable and efficient alternatives for the quantification of AA in pharmaceutical formulations, significantly reducing the environmental impact compared to conventional techniques. By incorporating ethanol and ultrapure water as greener solvents, the chromatographic and spectrophotometric methods demonstrated excellent analytical performance while adhering to the principles of green chemistry. These advancements offer safer, more eco-friendly solutions for pharmaceutical quality control, aligning with global efforts to minimize hazardous chemical usage and laboratory waste. Furthermore, the high AGREE scores achieved by the new methods highlight their environmental benefits, proving them to be valuable tools in advancing sustainability within analytical chemistry. As the demand for green methodologies continues to grow, these techniques offer a promising direction toward reducing the ecological footprint of pharmaceutical analysis without compromising accuracy or reliability, thereby contributing to safer laboratory practices and the broader goals of environmental preservation.

Funding: This research received no external funding.

Data Availability Statement: The data that support the findings of the study are available from the corresponding author upon reasonable request.

Conflicts of Interest: The author declares no conflict of interest.

References

1. Abdussalam-Mohammed, W.; Qasem Ali, A.; Errayes, A.O. Green Chemistry: Principles, Applications, and Disadvantages. *Chem. Methodol.* **2020**, *4*, 408–423. [CrossRef]
2. Kokilambigai, K.S.; Lakshmi, K.S. Analytical quality by design assisted RP-HPLC method for quantifying atorvastatin with green analytical chemistry perspective. *J. Chromatogr. Open* **2022**, *2*, 100052. [CrossRef]
3. Yabré, M.; Ferey, L.; Somé, I.T.; Gaudin, K. Greening Reversed-Phase Liquid Chromatography Methods Using Alternative Solvents for Pharmaceutical Analysis. *Molecules* **2018**, *23*, 1065. [CrossRef] [PubMed]
4. Snyder, L.R.; Kirkland, J.J.; Dolan, J.W. *Introduction to Modern Liquid Chromatography*; John Wiley & Sons: Hoboken, NJ, USA, 2011.
5. Welch, C.J.; Wu, N.; Biba, M.; Hartman, R.; Brkovic, T.; Gong, X.; Helmy, R.; Schafer, W.; Cuff, J.; Pirzada, Z.; et al. Greening analytical chromatography. *TrAC Trends Anal. Chem.* **2010**, *29*, 667–680. [CrossRef]
6. Guideline, I.H.T. Validation of analytical procedures: Text and methodology. Q2 (R1). *Curr. Step* **2005**, *1*, 5.
7. Sheldon, R.A. Fundamentals of green chemistry: Efficiency in reaction design. *Chem. Soc. Rev.* **2012**, *41*, 1437–1451. [CrossRef] [PubMed]
8. Jankech, T.; Gerhardtova, I.; Stefanik, O.; Chalova, P.; Jampilek, J.; Majerova, P.; Kovac, A.; Piestansky, J. Current green capillary electrophoresis and liquid chromatography methods for analysis of pharmaceutical and biomedical samples (2019–2023)—A review. *Anal. Chim. Acta* **2024**, *1323*, 342889. [CrossRef] [PubMed]
9. Płotka, J.; Tobiszewski, M.; Sulej, A.M.; Kupska, M.; Górecki, T.; Namieśnik, J. Green chromatography. *J. Chromatogr. A* **2013**, *1307*, 1–20. [CrossRef]
10. Shen, Y.; Chen, B.; van Beek, T.A. Alternative solvents can make preparative liquid chromatography greener. *Green Chem.* **2015**, *17*, 4073–4081. [CrossRef]
11. Yin, L.; Hu, Q. CYP17 inhibitors—Abiraterone, C17,20-lyase inhibitors and multi-targeting agents. *Nat. Rev. Urol.* **2014**, *11*, 32–42. [CrossRef]
12. Sonpavde, G.; Attard, G.; Bellmunt, J.; Mason, M.D.; Malavaud, B.; Tombal, B.; Sternberg, C.N. The Role of Abiraterone Acetate in the Management of Prostate Cancer: A Critical Analysis of the Literature. *Eur. Urol.* **2011**, *60*, 270–278. [CrossRef] [PubMed]

13. Neil, J.H. Methanesulfonate Salts of Abiraterone-3-Esters and Recovery of Salts of Abirater One-3-Esters from Solution in Meth Tert-Butyl Ether. EP1789432B1, 30 June 2010.
14. Acharya, M.; Bernard, A.; Gonzalez, M.; Jiao, J.; De Vries, R.; Tran, N. Open-label, phase I, pharmacokinetic studies of abirateron acetate in healthy men. *Cancer Chemother. Pharmacol.* **2012**, *69*, 1583–1590. [CrossRef]
15. Anonymous. Abiraterone: Uses, Interactions, Mechanism of Action | DrugBank Online. Available online: https://go.drugban com/drugs/DB05812 (accessed on 24 March 2024).
16. Bono, J.S.d.; Logothetis, C.J.; Molina, A.; Fizazi, K.; North, S.; Chu, L.; Chi, K.N.; Jones, R.J.; Goodman, O.B.; Saad, F.; et Abiraterone and Increased Survival in Metastatic Prostate Cancer. *N. Engl. J. Med.* **2011**, *364*, 1995–2005. [CrossRef] [PubMed]
17. Martins, V.; Asad, Y.; Wilsher, N.; Raynaud, F. A validated liquid chromatographic-tandem mass spectroscopy method for th quantification of abiraterone acetate and abiraterone in human plasma. *J. Chromatogr. B Anal. Technol. Biomed. Life Sci.* **200** *843*, 262–267. [CrossRef]
18. Kumar, S.V.; Rudresha, G.; Gurav, S.; Zainuddin, M.; Dewang, P.; Kethiri, R.R.; Rajagopal, S.; Mullangi, R. Validated RP-HPLC/U method for the quantitation of abiraterone in rat plasma and its application to a pharmacokinetic study in rats. *Biomed. Chromatog* **2013**, *27*, 203–207. [CrossRef]
19. Gurav, S.; Punde, R.; Farooqui, J.; Zainuddin, M.; Rajagopal, S.; Mullangi, R. Development and validation of a highly sensitiv method for the determination of abiraterone in rat and human plasma by LC-MS/MS-ESI: Application to a pharmacokinet study. *Biomed. Chromatogr.* **2012**, *26*, 761–768. [CrossRef]
20. Khedr, A.; Darwish, I.; Bamane, F. Analysis of abiraterone stress degradation behavior using liquid chromatography coupled t ultraviolet detection and electrospray ionization mass spectrometry. *J. Pharm. Biomed. Anal.* **2013**, *74*, 77–82. [CrossRef]
21. Wani, T.A. Highly sensitive ultra-performance liquid chromatography–tandem mass spectrometry method for the determinatio of abiraterone in human plasma. *Anal. Methods* **2013**, *5*, 3693–3699. [CrossRef]
22. Belleville, T.; Noé, G.; Huillard, O.; Thomas-Schoemann, A.; Vidal, M.; Goldwasser, F.; Alexandre, J.; Blanchet, B. A HPLC fluorescence method for the quantification of abiraterone in plasma from patients with metastatic castration-resistant prostat cancer. *J. Chromatogr. B* **2015**, *989*, 86–90. [CrossRef]
23. Gong, A.; Zhu, X. β-cyclodextrin Sensitized Spectrofluorimetry for the Determination of Abiraterone Acetate and Abiraterone *J. Fluoresc.* **2013**, *23*, 1279–1286. [CrossRef]
24. Annapurna, M.M.; Pradhan, D.P.; Chaitanya, R.K. New Spectrophotometric Techniques for the determination of Abiraterone acetate (A Steroidal Inhibitor) in Tablets. *J. Chem. Pharm. Sci.* **2017**, *10*, 768–770.
25. Burgess, C.; McDowall, R. Quo Vadis Analytical Procedure Development and Validation? *Spectroscopy* **2022**, *37*, 8–14. [CrossRef
26. Zhao, M.; Markiewicz-Keszycka, M.; Beattie, R.J.; Casado-Gavalda, M.P.; Cama-Moncunill, X.; O'Donnell, C.P.; Cullen, P.J Sullivan, C. Quantification of calcium in infant formula using laser-induced breakdown spectroscopy (LIBS), Fourier transform mid-infrared (FT-IR) and Raman spectroscopy combined with chemometrics including data fusion. *Food Chem.* **2020**, *320*, 12663 [CrossRef] [PubMed]
27. Shetty, S.; Shah, A. New spectrophotometric method for estimation of tolterodine in bulk and pharmaceutical formulation. *Int. J Pharm. Sci. Res.* **2011**, *2*, 1455.
28. Yin, L.; Yu, L.; Guo, Y.; Wang, C.; Ge, Y.; Zheng, X.; Zhang, N.; You, J.; Zhang, Y.; Shi, M. Green analytical chemistry metrics fo evaluating the greenness of analytical procedures. *J. Pharm. Anal.* **2024**, 101013. [CrossRef]
29. Cetinkaya, A.; Kaya, S.I.; Ozkan, S.A. An overview of the current progress in green analytical chemistry by evaluating recen studies using greenness assessment tools. *TrAC Trends Anal. Chem.* **2023**, *168*, 117330. [CrossRef]
30. Pena-Pereira, F.; Wojnowski, W.; Tobiszewski, M. AGREE-Analytical GREEnness Metric Approach and Software. *Anal. Chem* **2020**, *92*, 10076–10082. [CrossRef]
31. Bukkapatnam, V.; Mukthinuthalapati, M.A.; Routhu, K.C. Study on Stress Degradation Behaviour of Abiraterone Acetate in Film Coated Tablets and Identification of Major Stress Degradation Product by Liquid Chromatography-Ultraviolet-Electrospray Ionization-Mass Spectrometry. *Indian J. Pharm. Sci.* **2022**, *84*, 300. [CrossRef]
32. Mansour, F.R.; Omer, K.M.; Płotka-Wasylka, J. A total scoring system and software for complex modified GAPI (ComplexMoGAPI) application in the assessment of method greenness. *Green Anal. Chem.* **2024**, *10*, 100126. [CrossRef]

separations

MDPI

Review

Utilization and Separation of Flavonoids in the Food and Medicine Industry: Current Status and Perspectives

Jianan Wang [1], Bin Wang [1], Chunli Chen [1], Jingzhou Dong [1,*] and Huafeng Zhang [2,*]

[1] School of Selenium Science and Engineering, Wuhan Polytechnic University, Wuhan 430048, China; 28256wjn@aliyun.com (J.W.); wb17853580312@163.com (B.W.); chunlichen1@aliyun.com (C.C.)
[2] College of Food Engineering and Nutritional Science, Shaanxi Normal University, Xi'an 710062, China
* Correspondence: djz21cn@aliyun.com (J.D.); ijrc@snnu.edu.cn (H.Z.)

Abstract: Flavonoids are the most abundant functional compounds distributed in higher plants, and are used as important dietary components for human health protection. The development of natural flavonoids, such as functional food or medicinal food, has received extensive attention in recent years. The extraction, separation, and quantitation of flavonoids are the key techniques in the utilization of flavonoid resources. The traditional methods for flavonoid extraction and separation always used toxic solvents, which produce toxic residues and pollute the environment. Based on an analysis of the literature on flavonoid resources, the utilization, separation, quantitation, and green separation techniques of flavonoids were summarized. First, extraction by hot water or hot ethanol, assisted by pressurization and microwave-ultrasonication, then concentration and precipitation of flavonoids by cool water or cool ethanol or ethanol/water in specific ratios. This method could obtain over 85% purity in the first cycle and over 95% purity after three precipitation cycles in the separation of the most commonly used flavonoids, such as dihydromyricetin, rutin, and quercetin. In conclusion, flavonoids showed great prospects in human health protection and disease treatment. Chemical structure-based separation using the water–ethanol methods and assisted with microwave-ultrasonication, pressurization, and temperature regulation proved to be efficient and environmentally friendly, showing great potential for the flavonoid industry. These "green" processing techniques and mechanisms deserve further research.

Keywords: flavonoid resources; medicinal values; separation techniques; quantitation

1. Introduction

Natural flavonoids are the most abundant polyphenols in plant species. Over 15,000 flavonoids have been separated and identified from plants [1–6]. In plant kingdoms, flavonoids are one of the most important groups of plant secondary metabolites and are biosynthesized via a shikimate pathway, which is followed by a number of branch pathways leading to isoflavonoids, flavonols, proanthocyanidins, and anthocyanins [7].

Flavonoids such as rutin, quercetin, kaempferol, baicalin, luteolin, hesperetin, scutellarin, puerarin, dihydromyricetin, and myricetin were important compositions of traditional medicines for human disease treatment. Rutin and quercetin have been reported to have many bioactivities, such as antibacterial [8], anti-inflammation [9] and antivirus [10–13], but promoted the growth of *Aspergillus niger* [14]. Baicalin was also well known for its antivirus [15–17] and antibacterial [18,19] bioactivities, and was widely used as a medicinal injection "Qingkailing" in China. Luteolin showed antivirus activities [20], although it was the main component in traditional medicines for cardiovascular and cerebrovascular disease treatment. Dihydromyricetin, an important natural flavononol mostly existing in vine tea (*Ampelopsis grossedentata* leaves), with a high content of 16% [12] and traditionally used for anticold, antifaucitis, liver protection, and cardiovascular diseases treatment in China, showed strong antibacterial activities as well [21,22]. Nowadays, flavonoids such as dihydromyricetin, rutin, baicalin, and many other flavonoids are being developed or used as food preservatives and clinical drugs in tablets, capsules or injections. Since flavonoids are widely used as functional foods or medicines, investigations and studies on the resources of flavonoid-rich plant species and the related separation methods of the flavonoids from plants are crucial for the utilization of flavonoids.

The objective of this review is to investigate the flavonoid resources, classifications, medicinal values, and extraction and separation methods, which were published or industrially used in recent years, and finally, summarize the environmentally friendly separation process for the flavonoids industry.

2. Materials and Methods

The literature database includes Springer link, Science direct, PubMed, Wiley Online Library, International Patents, Chinese traditional medicinal records, and Chinese science databases. A total of 125 articles on flavonoid resources and chemical structures, medicinal utilizations, separation and quantitation methods, and the separation techniques of typical flavonoids, such as dihydromericitin, rutin, and qercetin, were investigated and summarized.

Flavonoid resources including plant species, distributions, chemical classifications, and utilizations are summarized and comparatively described in tables. Recent years' reviews on the medicinal properties and related mechanisms of flavonoids are summarized and described in tables. Based on analyzing the correlations between flavonoid chemical structures and separation techniques, the "green" and highly efficient separation techniques were suggested. The structure-based quantitation methods were comparatively analyzed and suitable determination methods for different utilizations of flavonoids were suggested.

3. Results and Discussion

3.1. Chemical and BiologicalProperties of Flavonoids

Flavonoids are a large number of small molecules with the similar structure of a 15-carbon skeleton-2-phenyl-benzopyranonee, which has a three-carbon bridge cyclized with oxygen between the phenyl groups (Figure 1). Based on their chemical structures, the most common flavonoids are classified as flavones, flavonols, flavanones, flavanonols, chalcones, isoflavones, isoflavanones, flavan-3-ol, flavan-3,4-diols, aurones, xanthones, and anthocyanidins. They are differentiated according to the degree of unsaturation and oxidation of the three-carbon segment [23]. Further differentiation is based on the number and nature of substituent groups attached to the rings [24]. Flavonoids can exist both as free aglycones and as glycosidic conjugates in plants [23,25]. In general, glycosylation makes flavonoids less reactive, more soluble in water. The glycosylation modifications are considered to be a protective form to prevent cytoplasmic damage, and to store the flavonoids safely in cell vacuoles [26]. For the glycosylated flavonoids, glucose is the most common sugar, followed by galactose, rahmnose, xylose, and arabinose [23]. Rutinose(6-O-α-L-rhamnosyl-D-glucose) and neohesperidose (2-O-α-L-rhamnosyl-D-glucose) are the most common disaccharides [24]. One or more hydroxyls of glycosides are occasionally acylated by groups such as acetic acid, ferulic acid, or caffeic acid [23,27].

Much of the published literature indicated that flavonoids might be involved in disease resistance in plants and act as "phytoalexin". Most flavonoid phytoalexins have been isolated from underground parts, such as roots, rhizomes, hypocotyls, tubers or seeds [28].The most important biological functions of flavonoids could be summarized as ultraviolet (UV) filtration and free radical scavenging [29]. The natural flavonoids kaempferol, quercetin, and rutin were considered as photoprotectors, as they could absorb the main UV radiation within 200–400 nm. This is why many flavonoids such as rutin and quercetin are widely used as the main components of sunscreens. In fact, almost all flavonoids showed both UV absorption and radical scavenging functions. It is well known that plants distributed at higher altitudes contain higher contents of flavonoids, especially purple potatoes [30], since plants at higher altitude receive more UV radiation, which is the promoter for plant flavonoid accumulation. Free radical scavenging activity (antioxidation), which is the best-described property of flavonoids, might be the basic mechanism for many pharmacological activities of flavonoids. This basic mechanism is that flavonoids have lower redox potentials to reduce higher oxidizing free radicals by forming less-reactive radicals, which are much less harmful than the free radicals, and accordingly, prevent many oxidation damages, such as lipid peroxidation and cellular membrane damage, which leads to cell death. Take, for example, pelargonidin, which seems more likely to derive from their strong radical scavenging action.

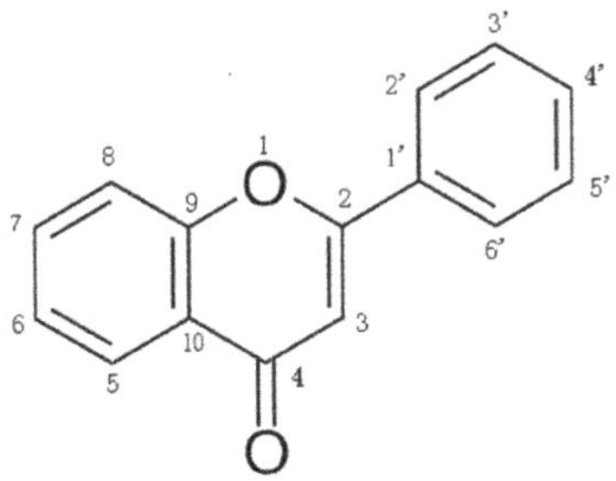

Figure 1. Basic structure of flavonoids: 2-phenyl-chromone.

3.2. *Resources and Utilizations of Flavonoids*

Natural products of plants are being considered as new sources of potential drugs or functional foods. Among various phytochemicals, flavonoids have received increasing attention due to their considerable bioactivities for human health protection. Flavonoids are the largest group of naturally occurring phenolic compounds, which occur commonly in different parts of higher plants (roots, tubers, bark, heartwood, leaves, flowers, fruits, and seeds) both in a free state and as glycosides [23]. Flavonoids have been reported as the major dietary constituents of plant foods [31]. It has been estimated that 3–80 mg of flavonoids should be consumed daily [32,33]. Since flavonoids had been reported to have the basic bioactivity of free radical scavenging, as summarized above, many plant species rich in flavonoids are widely used as medicinal foods or functional foods in China, Southeast Asia, and many other countries. Many plant species containing high contents of flavonoids have been widely used as either traditional medicines or functional foods, mostly in the form of teas and vegetables, from ancient times. Many kinds of tea such as green tea (leaves of *Camellia sinensis*), vine tea (leaves of *Ampelopsis grossedentata*), chrysanthemum tea (flowers of *Dendranthema morifolium*), wolfberry tea (buds of *Lycium barbarum* or leaves of *L. barbarum* and *L. chinense*), and sophora tea (flowerbuds of *Sophora japonica*) have been used as medicine, food and homological drinks for health improvement and chronic disease treatment. For their conveniences in packaging, storage and use, these teas were considered the best functional foods or medicinal foods by human beings. The main medicinal values and utilizations of most widely used flavonoid plant species are summarized in Table 1.

Table 1. Plant resources and medicinal uses of flavonoids.

Main Plant Species	Main Flavonoids	Distribution Parts	Utilizations	Main Medicinal Values	Cites
Sophora japonica	Rutin	Fruits, buds	Tea, TM	Cochlear protection	[34]
Fagopyrum esculentum	Rutin	Leaves	Tea, TM	Cochlear protection	[34]
Ampelopsis grossedentata	Dihydromyricetin	Leaves, fruits	Tea, TM	Anti-inflammation	[35]
Apium graveolens	Apigenin	Leaves, stems	Food	Liver protection	[36]
Kaempferia galanga	Kaempferol	Rhizomes	TM	Anti-cancer	[37]
Sophora flavescens	Quercetin	Buds	TM	Anti-cancer	[38]
Myrica rubra	Myricetin	Leaves, roots, barks	Food, TM	Anti-cancer	[39]
Trifolium pratense	Daidzein	Entire plants	Feed, TM	Anti-cancer	[40]

Table 1. *Cont.*

Main Plant Species	Main Flavonoids	Distribution Parts	Utilizations	Main Medicinal Values	Cites
Medicago sativa	Daidzein	Entire plants	Food, feed	Anti-cancer	[41]
Glycine max	Daidzein	Seeds	Food	Anti-cancer	[41]
Oroxylum indicum	Chrysin	Seeds, barks	TM	Anti-cancer; Anti-inflammation	[42,43]
Scutellaria baicalensis	Baicalein	Entire plants	TM	Apoptosis induction	[44]
Alpinia officinarum	Galangin	Roots	TM	Anti-inflammation;anti-proliferation	[45,46]
Rhus succedanea	Fisetin	Leaves, stems	TM	Anti-inflammation	[47]
Hibiscus sabdariffa	Gossypetin	Flowers	Tea, TM	Anti-oxidation	[48]
Genista tinctoria	Genistein	Hizome	TM	Neuroprotection	[49]
Turnera diffusa	Pinocembrin	Leaves	TM	Neuroprotection	[50,51]
Prunus yedoensis	Naringenin	Buds	TM	Neuroprotection	[52]
Eriodictyon californicum	Eriodictyol	Entire plants	TM	Anti-cancer	[53]
Glycine max	Cyanidin	Black episperm	Food, TM	Neuroprotection	[54]
Lonicera japonica	Luteolin	Flowers, buds	Tea, TM	Anti-oxidation;neuroprotection	[55,56]
Eucalyptus globulus	Tricetin	Leaves	TM	Anti-breast cancer	[57]

TM = traditional medicine.

In recent years, many review articles have summarized the bioactivities of natural flavonoids in human disease prevention and related mechanisms. Most of them focused on anti-diabetes, anti-oxidation, neuroprotection, antivirus, anti-inflammation, anticancer, liver protection, and so on. The main reviews on the pharmacological values and related mechanisms of flavonoids are summarized in Table 2.

Table 2. Main coverage of review articles on flavonoid medicinal values in recent years.

Medicinal Values	Flavonoids	Possible Mechanisms	Cites
Treatment of hepatitis.	Quercetin, baicalin, silymarin	Protective effects on acute and chronic hepatitis, hepatitis C, fatty liver and toxic liver injury in in vitro experiments and clinical trials.	[58]
Anti-diabetes	Quercetin, baicalin, kaempferol, myricetin, rutin, apigenin, naringenin, hesperidin, genistein, proanthocyanidin	Suppressing the upstream oxidative stress and inflammatory pathways to promote podocyte autophagy and inhibit the overactivity of RAAS, ultimately alleviate DN, blocking the enzymes involved in insulin resistance, facilitating glucose metabolism, and positively affecting plasma glucose and insulin levels.	[59,60]
Anti-oxidation	Rutin, baicalein, apigenin, quercetin	Serving as a natural antioxidant, antiviral, anti-inflammatory, cardioprotective, anti-diabetic, anti-cancer, anti-aging, and other bioactive qualities.	[61]
Antifungal activity	Proanthocyanidin, polymeric tannins, epigallocatechin-3-gallate, loureirin A epigallocatechin, epicatechin-3-gallate, lichochalcone-A	Biofilm disruption or inhibition, along with other cell inhibitory activities.	[62]
Antiangiogenisis	Flavone, Wogonin, wogonoside, flavonoids 3-hydroxy flavone, hesperidin, apigenin, fisetin	Inhibiting the proliferation and migration of endothelial cells by interfering with key angiogenic signal cascades, such as MAPK and PI3K pathways.	[63]
Programmed cell death induction	Quercetin, tangeretin, myricetin, naringin, pectolinaringenin, apigenin, kaempferol, chrysin	Activating different forms of cell death by regulating key pathways.	[64]
Antivirus(COVID-19), anti-inflammation	Kaempferol, quercetin	Inhibiting protein kinase B and phosphorylation of protein kinase and blocking effects on 3a channel; reducing the level of ROS, expression of iNOS, and pro-inflammatory mediators.	[65]
Cardioprotection	Kaempferol, apigenin, anthocyanins, quercetin, rutin, apigenin	Antioxidation, anti-atherosclerosis and anti-thrombosis, and heart protection.	[66]
Pulmonary arterial hypertension treatment	Chrysin, isorhamnetin, kaempferol, hesperidin, pinocembrin, puerarin, hydroxysafflor, yellow A, dihydromyricetin, catechin	Inhibiting inflammation, oxidative stress, ion channels, and cell proliferation to slow down the occurrence and development of PAH.	[67]

Table 2. *Cont.*

Medicinal Values	Flavonoids	Possible Mechanisms	Cites
Anti-cancer	Daidzein, hesperetin, cocoa catechins, chrysin, cyanidin, vitexin, baicalein, isoliquiritigenin	Inducing apoptosis, necrosis, suppressing migration, arresting cell cycle phases, modulating signaling pathways.	[68–70]
Anti-neuroinflammation	Hesperidin, apigenin, morin, naringin, rutin, epigallocatechin-3-gallate, Icariside II, anthocyanins, quercetin, wogonin	Regulating enzyme and cell function, and preventing neurodegenerative diseases through various signal transduction pathways.	[71]
Skeletal Muscle Health promotion	Resveratrol, (-)-epicatechin, dihydromyricetin, apigenin	Modulating molecular pathways involved in the development of mitochondrial myopathy to act as a muscle regulator.	[72]
Anti-hypertension	Apigenin, chrysin, artemetin, fisetin, (+)-Catechin, genistein, naringin, anthocyanin, cardamonin	Affects hypertension in many ways and can be used as a supplement to the routine treatment of hypertension.	[73]
Anti-obesity	Chrysin, apigenin, quercetin, catechins, naringin, naringenin, daidzein, delphinidin	Reduces body weight, fat mass and plasma triglycerides/cholesterol by reducing food intake and fat absorption, increasing energy expenditure, modulating lipid metabolism, or regulating gut microbiota profile.	[74]
Viral acute respiratory tract infectionstreatment	Propolis, quercetin, naringenin, . troxerutin	Treating viral ARTIs including the common cold, influenza, COVID−19, acute bronchitis, bronchial pneumonia, and upper respiratory tract infections.	[75]
Depressive symptom improvement	Rutin, apigenin, glycitein, isoflavones, genistein, epigallocatechin-3-gallate	Interactions with oxidative pathways and antioxidant systems; modulating signaling pathways responsible for maintaining neuron survival and synaptic plasticity.	[76]
Anti-angiogenic effects	Genistein, biochanin-A, hesperidin, myricetin, catechin	Regulating angiogenesis to inhibit cancer, arthritis, and retinopathy.	[77]

3.3. Chemical Structure BasedClassification and Solubilities of Flavonoids

Besides the basic structure of 2-phenyl-chromone, the second essential feature of flavonoids is that they have hydroxyl groups, which usually decide their antioxidation activities and solubilities. As shown in Table 3, flavonoids with different numbers of hydroxyls show different polarities. Namely, the more hydroxyl groups, the more polarity. It is the number of hydroxyls in flavonoids that decides their polarities and related extraction methods. Flavonoids with more hydroxyls could be more easily dissolved in hot water or ethanol, such as dihydromyricetin, myricetin, quercetin, and quercetagetin. Flavonoid glycosides are more soluble in water than flavonoids [23]. It is shown in Table 3 that flavonoids with more than five hydroxyls could be easily dissolved in hot water, but slightly dissolved in cool water. Flavonoid glycosides such as rutin, baicalin, and icariin could be easily dissolved in boiled water. These chemical properties indicated the special extraction methods of flavonoids.

Table 3. Flavonoid families and their water–ethanol solubilities.

Chemical Structures	Names	Classification	Water Solubility	Ethanol Solubility	Optimized Solvents	References
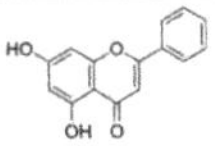	Chrysin	Flavone	NS	SS	80% ketone	[78,79]
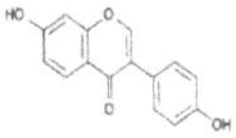	Daidzein	Isoflavone	NS	S	90% ethanol	[80]
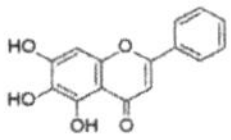	Baicalein	Flavone	SS	S	50%, 70% ethanol	[81–83]

Table 3. *Cont.*

Chemical Structures	Names	Classification	Water Solubility	Ethanol Solubility	Optimized Solvents	References
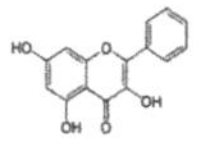	Galangin	Flavonol	SS	S	80% ethanol	[84,85]
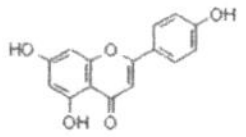	Apigenin	Flavone	SS	S	70%, 80% ethanol	[86,87]
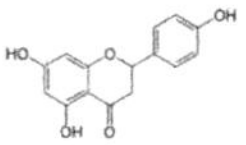	Naringenin	Flavanone	SS	S	70% ethanol	[88]
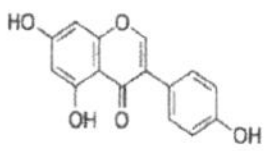	Genistein	Isoflavone	SS	S	66% ethanol	[89]
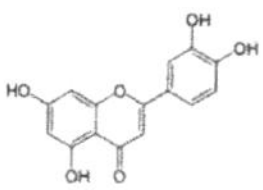	Luteolin	Flavone	SS	S	70%, 80% ethanol	[90,91]
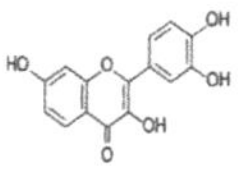	Fisetin	Flavonol	SS	S	65% ethanol	[92]
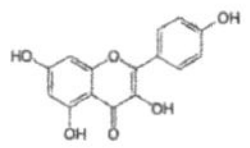	Kaempferol	Flavonol	SS	VS	70%, 95% ethanol	[93,94]
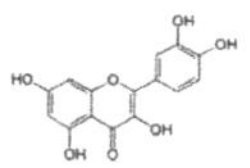	Quercetin	Flavonol	SS	VS	50%, 95% ethanol; 90 °C water	[93,95,96]
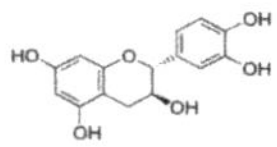	Catechin	Flavan-3-ol	▲VS	VS	80 °C water; 78% ethanol	[97,98]
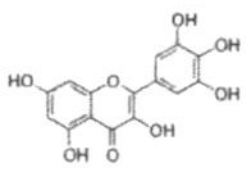	Myricetin	Flavonol	▲SS	VS	90% ethanol; 100 °C water	[99,100]
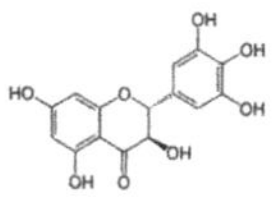	Dihydromyricetin	Flavanonol	▲SS	VS	90% ethanol; 100 °C water	[99,100]
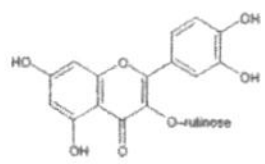	Rutin	Glycoside	▲SS	VS	65 °C water	[101]
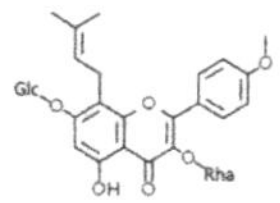	Icariin	Glycoside	▲SS	VS	100 °C water	[102]
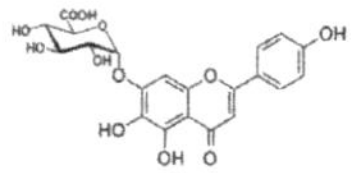	Baicalin	Glycoside	▲SS	VS	100 °C water	[103]

SS: slightly soluble; S: soluble; VS: very soluble; NS: not soluble; ▲: means SS in cool water (room temperature), but S or VS in hot water (95 °C).

3.4. ChemicalStructure-Based Extraction and Separation of Flavonoids

3.4.1. Extraction of Flavonoids: Traditional Methods and "Green" Methods

In recent decades, numerous extraction methods have been reported on the isolation of flavonoids from various plant materials. These traditional extraction methods of flavonoids from plant powders could be categorized as direct extraction by solvents [104], alkali extraction, acid precipitation [104,105], lead acetate precipitation [104], and supercritical fluid (liquid CO_2) extraction [106]. Alkali extraction, acid precipitation and lead acetate precipitation produce massive water waste and pollute environments, and could not be used in industrial extraction. Extraction with supercritical fluid (liquid CO_2) was of a high cost and could not be used on a large scale, although it was environmentally friendly. Hence, selection of the suitable liquid solvents for extraction of flavonoids was essential for the flavonoid industry. Organic solvents such as ethanol, methanol, and ethyl acetate [107,108] were widely used in the extraction of flavonoids. In the traditional extraction process, toxic solvents, such as petroleum ether and chloroform, were usually used for de-oiling treatment, before flavonoid extraction [104]. Solvents such as methanol, ethyl acetate, petroleum ether, and chloroform always produced solvent residues and environmental pollution, and were regarded as not in compliance with good manufacturing practice (GMP). In all these extraction solvents, pure water and food-grade ethanol are considered safe and environmentally friendly, so the suitable extraction of flavonoids with water or ethanol should be developed for industrial application.

Ultrasonics and microwaves were widely used for the assisted extraction of flavonoids, both on a laboratory level or an industrial scale, since both ultrasonics and microwaves could destroy the wall of plant cells [109,110]. Ultrasonication could accelerate mass transformation and promote the penetration of plant cell walls, ultrasonically induced cavitation, and vibration facilitating the swelling of the cells, causing cell fragmentation [110]. While microwaves generated fractures on cell walls due to the sudden intracellular temperature rise, internal pressure increased and resulted in the consequent detonation of cells [111]. So, in order to utilize the effects of ultrasonics and microwaves, a combination of ultrasonics and microwaves could be applied in the extraction of flavonoids [110]. Pressurization, as a "green" process, could promote cell rupture and the unfolding of cell membranes, which causes the increase in the extraction rate of flavonoids [112]. So, water–ethanol extraction assisted by microwaves, ultrasonication or pressurization could be developed as the "green" process for flavonoid extractions.

3.4.2. Separation of Flavonoids: "Green" Methods

Since the de-oiling treatment with toxic solvents in traditional extraction of flavonoids showed disadvantages, safe and practical extraction methods of flavonoids deserve further research. There have been some new methods reported for flavonoid extraction, which were economical, practical, and safe. As shown in Figure 2, in the traditional extraction methods, plant powders were first prepared by a de-oiling treatment with chloroform or petroleum ether, and then extracted with organic solvents such as methanol or ethyl acetate, which produces toxic residues in flavonoid extracts. Most of the flavonoids including aglycones and flavonoid glycosides could be easily dissolved in hot water and slightly dissolved in cool water (Table 3), which indicated new methods for extraction and separation of flavonoids.

Take dihydromyricetin separation as an example (Figure 2) [12,99,100]. Dihydromyricetin and other components, including polysaccharides, starches, proteins, chlorophylls, and other fatty compounds were first pressure-extracted using 90–100 °C water at 1.0 MPa, then, the hot water extracts were concentrated and treated with precipitation at a low temperature (0–10 °C). The precipitates were separated, dried, ground, and re-extracted by ethanol (95%, v/v).In this ethanol extraction process, dihydromyricetin was separated from the water-soluble impurities, such as proteins, starches, and polysaccharides, and in the vacuum concentration process of the ethanol extracts at 60 °C, fat-soluble compounds were precipitated first (Figure 3). UV-visible spectra of the precipitates from the concentration of ethanol extracts showed no absorption peak (290 nm) of dihydromyricetin (Figure 3), which meant that dihydromyricetin was well isolated from these fat-soluble compounds. Finally, the concentrated ethanol extracts were gradually mixed into cool water (0 °C) in the ratio of 1:20 (ethanol extracts/cool water=1/20), mL/mL to slowly precipitate dihydromyricetin (Figure 4). In this gradual precipitation, dihydromyricetin was further separated from the water-soluble polyphenols, and so were the fatty compounds such as chlorophylls or other oils (Figure 4). With this extraction method, dihydromyricetin of 85% purity was obtained in the first cycle, and after three such separating cycles, highly purified dihydromyricetin could be obtained (Figure 5), with purity over 95%, while the 5% impurities were mostly polyphenols, which were co-precipitated with dihydromyricetin by molecular absorption. Based on the published data [12,99,100], the methods for the separation of dihydromyricetin from vine tea (*Ampelopsis grossedentata*) were optimized as described above (Figures 2–5),and successfully used in the vine tea industry.

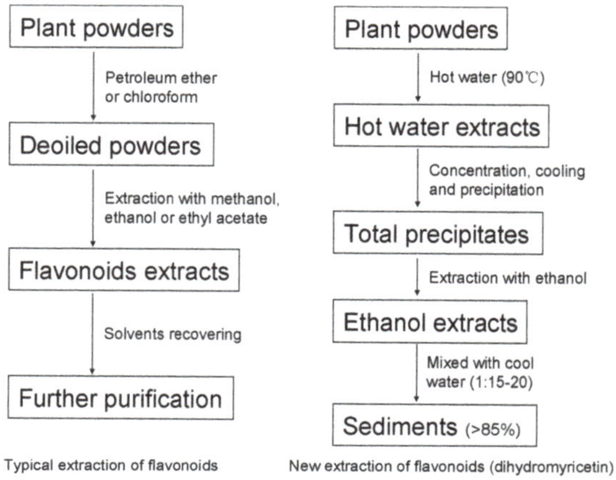

Figure 2. Comparison of flavonoid extraction between traditional methods and new methods [12,99].

Figure 3. Precipitates of ethanol extracts in concentration process (**left**) and UV-visible spectra of the precipitates(**right**) in absolute ethanol [12,100].

Figure 4. Precipitation purification process of dihydromyricetin (ethanol extracts/cool water = 1/20): A = chlorophylls and other lipids; B = water soluble polyphenols; and C = dihydromyricetin precipitates (>85% in the first cycle, >95% after three cycles) [12,99].

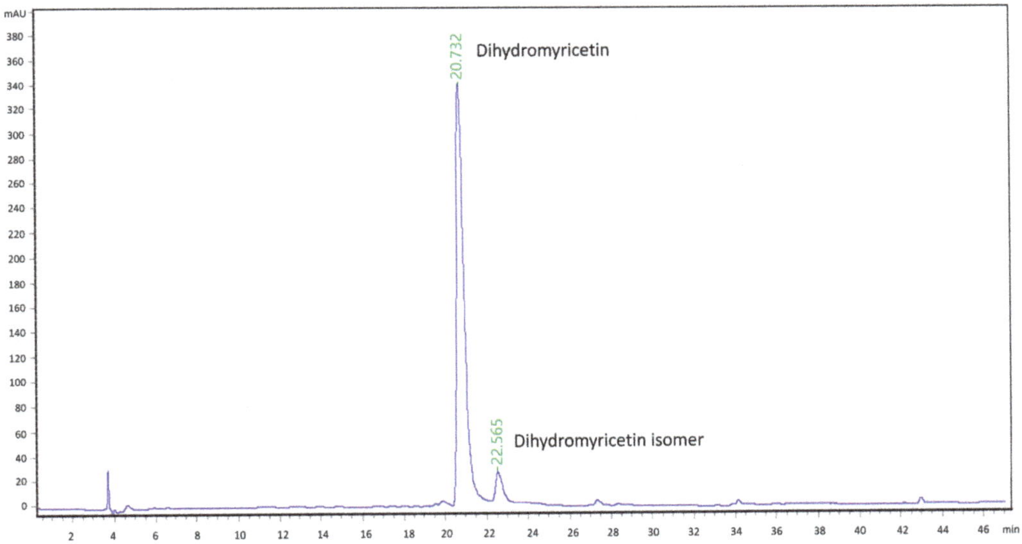

Figure 5. HPLC separation of the flavonoids from part C of Figure 4 (HPLC system: Hangzhou Saizhi Sci & Tech Co., Ltd., Hangzhou, China, N2000ChemStation, STI501 pump, STI 501,Lichrospher C18 column, 4.6 × 250 mm, Jiangsu Hanbon Sci &Tech Co., Ltd., Nanjing, China); DAD detection wavelengths: 220–500nm; mobile phase: acetonitrileas solvent A (1.0% acetic acid as solvent B) 0–25 min, 10–25%, 25–28 min, 25–90%, 28–30 min, 90–10%. The column temperature was 25 °C, sample volume injection 20.0 μL, flow rate1.0 mL/min [12].

Scientific methods to isolate the impurities such as chlorophylls, lipids, and other water-soluble compounds such as polyphenols, sugars, organic acids, amino acids, etc., from flavonoid extracts were the key techniques in the separation of flavonoids. Based on their chemical structures and related properties, flavonoids of plant powders could be efficiently extracted and separated with "green" solvents such as water and ethanol, rather than poisonous solvents such as methanol, ethyl acetate, petroleum ether, and chloroform.

Although supercritical fluid extraction by liquid CO_2 was considered the most environmentally friendly, the total flavonoid extracts by liquid CO_2 still needed to be separated and purified. Nowadays, solid phase separation methods such as column chromatography are widely used for flavonoid industrial separation. However, the column chromatography process produces a great amount of waste solvents. So, the environmentally friendly separation of flavonoids is crucial for the flavonoid industry. The water–ethanol separation methods described above showed great potential in the industrial separation of flavonoids. The water–ethanol separation techniques assisted with ultrasonication, microwaves, temperature regulation, and pressurization, and deserve further research.

Interestingly, two or more flavonoids might be combined to exert a synergistic effect when they are used as functional food ingredients or phytomedicines in a mixture [113,114]. In this sense, it might be better to use a concentrated mixture of flavonoids in foods. Namely, we could only extract and separate total flavonoids from plant materials, and there was no need to separate individual components in the food industry. In flavonoid medicines, combinations of individual flavonoids and their synergic pharmacological effects deserve further study too.

3.5. *Quantitation of Flavonoids*

3.5.1. Sample Preparation

The quantitation of flavonoids is very important in the separation of flavonoids. Sample preparation is the first step for flavonoid quantitation. Sample preparation depends on the nature of analytes (e.g., specific flavonoids subclasses or total polyphenolic compounds, aglucones or glycosides) and samples (e.g., physical state). Samples such as plant powders, foods, and liquids (fruit juices) are usually treated separately. The solid samples and fresh fruits are usually dried and ground into powders and extracted with suitable solvents, while liquid samples such as wine, drinks or fruit juices are always isolated by liquid–liquid extraction or solid phase extraction (Figure 6).

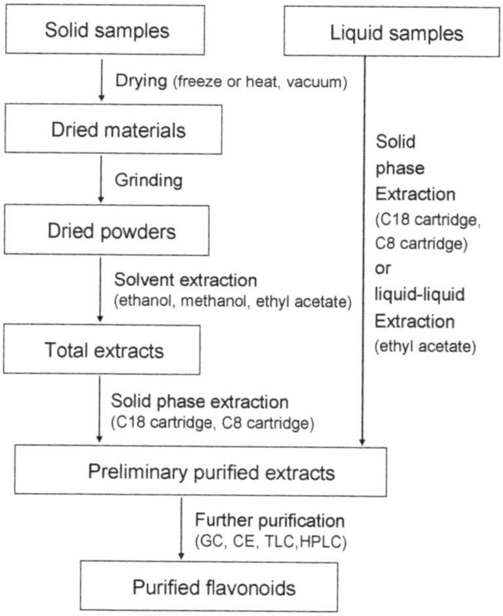

Figure 6. General sample preparation for flavonoid quantitation.

Sometimes, in order to determine the target analytes in aglycones, preliminary hydrolysis by acid or enzymes must be performed in the sample preparation. Acid hydrolysis (hydrochloric acid or formic acid) in the presence of ethanol was the traditional hydrolysis approach for preparation of flavonoid aglycones [115]. In the solvent extraction of flavan-3-ols (catechins, proanthocyanidins), aqueous acidic solvents should be used since the flavan-3-ols were unstable in neutral or alkaline environment [115].

Solid phase extraction is always used for the preliminary isolation of flavonoids from interfering compounds such as sugars. Minicartridges with C_{18}-bonded silica or C_8-bonded silica are widely used for the preliminary fractionation of flavonoid samples. With these cartridges, water-soluble compounds such as sugars can be eluted with methanol prior to the elution of phenols. The retained flavonoids are always eluted with methanol–dimethyl sulphoxide, which can enhance the solubility of flavonoids.

3.5.2. Quantitation Methods

There are various analytical methods for flavonoids. These analytical methods include thin layer chromatography (TLC), the spectrophotometric method, high-performance capillary electrophoresis (HPCE), high-performance liquid chromatography (HPLC), and liquid chromatography-mass spectrometry (LC-MS). These analytical methods showed advantages and disadvantages, respectively. In the food industry, suitable methods should be selected for the quantitation of extracted flavonoids.

Spectrophotometric method, colorimetric analysis: in most plant materials, the amount of individual flavonoids in plants is usually low, although some flavonoids such as dihydromyricetin and rutin were the main flavonoid compositions in leaves of *Ampelopsis grossedentata* and buds of *Sophora japonica*, respectively [109]. In most cases, flavonoids in plant materials are considered "total phenolics". Much of the published literature has reported that total phenolics could be conveniently quantified by spectrophotometric measurement through a simple extraction of the plant materials. This spectrophotometric measurement depends on colorimetric reactions, in which aluminum irons complexes with a pair of adjacent hydroxyls or with groups of 5-OH and 4-oxo, and then a strong absorption peak within 480–530nm of the complexes can be used for spectrophotometric analysis. Folin–Ciocalteu is the most commonly used colorimetric method for flavonoid quantitation. However, this spectrophotometric method always provides higher contents of flavonoids than the actual ones, since some unknown interfering components with similar groups of flavonoids take part in the complexation reaction.

Spectrophotometric method, direct quantitation: all flavonoid aglycones contain at least two aromatic rings and consequently absorb UV light of 240–285 nm (bandI) and 300–560nm (band II). The substitution substituents such as methyl, methoxy, and sugars generally produced minor changes in the absorption peaks [116]. Sometimes, there are only one or two flavonoids in a separated flavonoid extract, which can be quantified by the direct spectrophotometric method. In this direct spectrophotometric method, selecting suitable detection wavelength and standards was crucial [117]. However, no single wavelength was ideal for all classes of flavonoids, since they displayed absorbance maxima at distinctly different wavelengths. It has been reported that the total flavonoids of extracts could be quantified by a direct spectrophotometric method with rutin as the standard [117–119]. In order to determine flavonoids from extracts more accurately, dual-wavelength spectrophotometric methods have been used [120–124]. The multi-wavelength spectrophotometry could maximally reduce the interference of other compounds and significantly increase the accuracy of determination of the main flavonoids from extracts.

High-performance liquid chromatography (HPLC) method: HPLC was most widely used for the accurate determination of individual flavonoid components. Traditional HPLC used only one wavelength for detection of analytes and so could not conveniently detect all the flavonoids of samples [93,110,115]. HPLC with a diode array detector (DAD) could provide multi-wavelength detection, which could detect almost all flavonoids [122–124]. DAD-HPLC has become the necessary instrument in a standard analysis lab. HPLC combined with mass spectrometry (MS) provided convenience for the quantitation and identification of flavonoids. With the help of HPLC-MS, GC-MS, and UV detection, unknown flavonoids could be possibly identified.

4. Conclusions

Flavonoids showed many important pharmacological functions for human health protection, and were used as either functional foods or medicines. Flavonoid-rich plants or extracted flavonoids are being developed as materials for medicinal foods or natural medicines. Flavonoids of various chemical structures showed different water solubilities. The green process is crucial for flavonoid separations. Pure water or food-grade ethanol with pressurization, microwave-ultrasonication, and temperature regulation could be developed as highly efficient "green" methods for the extraction and separation of flavonoids. Flavonoids from plant materials could first be extracted by hot water or hot ethanol, and further precipitated by water temperature regulation or water–ethanol ratio regulation. This "green" method is environmentally friendly and highly efficient, and might be a better method for flavonoid separation. In this hot water extraction process, pressurization, microwaves and ultrasonication could be used to increase extraction rate.

Although the green separation of flavonoids is environmentally friendly, energy-saving and highly efficient, there are still some limitations such as the development of equipment for pressurization, ultrasonication, microwave emission, and temperature regulation.

In conclusion, water–ethanol separation of flavonoids assisted with pressurization and microwave-ultrasonication should be the future process for flavonoid production, and deserves further research.

Author Contributions: J.W.: Conceptualisation; Methodology; Formal analysis; Data curation; Writing—original draft; Writing—review & editing; Visualisation. C.C.: Conceptualisation; Methodology; Writing—review & editing; Investigation, Visualisation. B.W.: Methodology; Writing—review & editing. J.D.: Conceptualisation; Methodology; Resource; Project administration; Supervision; Writing—original draft; H.Z.: Writing—review & editing; Visualisation; Funding acquisition. All authors have read and agreed to the published version of the manuscript.

Funding: This research was funded by the National Natural Science Foundation of China (Grant No. 30570172) and Doctoral Research Project of Wuhan Polytechnic University (Grant No. 53210052493).

Acknowledgments: We are grateful to JinqiTengcha Bio. Ltd., Laifeng (China) for providing experimental help and to anonymous reviewers and scientific editors for critical review and valuable suggestions.

Conflicts of Interest: The authors declare no conflict of interest.

References

1. Salazar, H.M.; Mendonça, R.D.; Laclaustra, M.; Moreno-Franco, B.; Åkesson, A.; Guallar-Castillón, P.; Donat-Vargas, C. The intake of flavonoids, stilbenes, and tyrosols, mainly consumed through red wine and virgin olive oil, is associated with lower carotid and femoral subclinical atherosclerosis and coronary calcium. *Eur. J. Nutr.* **2022**, *61*, 2697–2709. [CrossRef] [PubMed]
2. Wang, Y.L.; Zhao, Y.; Yang, F.; Yuan, Y.M.; Wang, H.; Xiao, J. Influences of Glucose on the Dietary Hydroxy flavonoid-Plasma Protein Interaction. *J. Agric. Food Chem.* **2012**, *60*, 12116–12121. [CrossRef] [PubMed]
3. Li, M.; Hagerman, A.E. Interactions Between Plasma Proteins and Naturally Occurring Polyphenols. *Curr. Drug Metab.* **2013**, *14* 432–445. [CrossRef]
4. Veitch, N.C.; Grayer, R.J. Flavonoids and their glycosides, including anthocyanins. *Nat. Prod. Rep.* **2011**, *28*, 1626–1695. [CrossRef]
5. Taneja, I.; Arora, S.; Rama Raju, K.S.; Siddiqui, N. Disposition of Pharmacologically Active Dietary Isoflavones in Biological Systems. *Curr. Drug Metab.* **2013**, *14*, 369–380.
6. Xiao, J.B.; Xie, Y.X.; Wang, Y.L.; Cao, H.; Chen, X.Q. Glycation of Plasma Proteins in Diabetics Will Reduce Benefits of Antioxidant Capacity of Natural Polyphenols. *Free Radic. Biol. Med.* **2012**, *53*, S90. [CrossRef]
7. Robbins, R.J. Phenolic acids in foods: An overview of analytical methodology. *J. Agric. Food Chem.* **2003**, *51*, 2866–2887. [CrossRef]
8. Ivanov, M.; Novovic, K.; Malesevic, M.; Dinic, M.; Stojkovic, D.; Jovcic, B.; Sokovic, M. Polyphenols as Inhibitors of Antibiotic Resistant Bacteria-Mechanisms Underlying Rutin Interference with Bacterial Virulence. *Pharmaceuticals* **2022**, *15*, 385. [CrossRef]
9. Wang, H.Y.; Wang, L.; Guo, S.S.; Liu, Z.H.; Zhao, L.Q.; Qiao, R.Z.; Li, C. Rutin-Loaded Stimuli-Responsive Hydrogel for Anti-Inflammation. *ACS Appl. Mater. Interfaces* **2022**, *14*, 26327–26337. [CrossRef]
10. Wang, Y.F.; Wang, X.H.; Zhu, Y.T.; Sun, H.D. Study on inhibitory effects of rutin on influenza A virus. *Chin. Arch. Tradit. Chin. Med.* **2005**, *23*, 827.
11. Chen, C.; Jiang, Z.Y.; Yu, B.; Wu, X.L.; Dai, C.Q.; Zhao, C.L.; Ju, D.H.; Chen, X.Y. Study on the anti-H1N1 virus effects of quercetin and oseltamivir and their mechanism related to TLR7 pathway. *J. Asian Nat. Prod. Res.* **2012**, *14*, 877–885. [CrossRef] [PubMed]

12. Zheng, X.J.; Xiao, H.; Zeng, Z. Composition and serum antioxidation of the main flavonoids from fermented vine tea (*Ampelopsis grossedentata*). *J. Funct. Foods* **2014**, *9*, 290–294. [CrossRef]
13. Nair, M.P.; Saiyed, Z.M.; Gandhi, N.H.; Ramchand, C. The flavonoid, quercetin, inhibits HIV-1 infection in normal peripheral blood mononuclear cells. *Am. J. Infect. Dis.* **2009**, *5*, 135–141. [CrossRef]
14. Liu, H.; Bian, J.S.; Huang, H.Y.; Yin, Z.W. Study on the Positive Effect of Rutin on the Growth of *Aspergillus niger*. *J. Xiangnan Univ.* **2013**, *34*, 121–124.
15. Zhang, C.J.; Yu, H.T. The signal pathways of immune inflammation mediated By The Tlr3/Nf-Kappab and activator protein-1 in cells infected with influenza A virus antagonized by Baicalin. *Adv. Mat. Res.* **2012**, *345*, 201–209. [CrossRef]
16. Chen, X.; Zhang, C.Y.; Gai, X.D. Study on Anti-Human Herpes Virus 6 Effects of Baicalin. *Appl. Mech. Mater.* **2012**, *140*, 48–52. [CrossRef]
17. Chen, X.; Zhang, C. Inhibitory role of baicalin on human herpes virus type 6 in vitro. *Procedia Eng.* **2012**, *37*, 75–78. [CrossRef]
18. Wang, Z.-Z.; Jia, Y.; Wang, G.; He, H.; Cao, L.; Shi, Y.; Miao, M.; Li, X.-M. Dynamic covalent hydrogel of natural product baicalin with antibacterial activities. *RSC Adv.* **2022**, *12*, 8737–8742. [CrossRef]
19. Zhu, M.; Li, L.; Zhang, T.L.; Bo, L.; Hao, W.L.; Chen, Z.B. Baicalin alleviates lung injury coused by *Staphylococcus aureus* pneumonia. *J. N. Pharma* **2014**, *11*, 65–67.
20. Gong, G.; Gong, G.; Qian, Z.; Zhou, S. Inhibitory activity of luteolin against transmissible gastroenteritis virus. *J. China Pharm. Univ.* **2005**, *36*, 453–456.
21. Wang, X.; Hu, H.; Hu, B.; Xia, H.; Cheng, X.; Zheng, J.; Zhang, Z.; Liu, H. Dihydromyricetin inhibits Hepatitis B virus replication by activating NF-κB, MAPKs, and autophagy in HepG2. 2.15 cells. *Mol. Biol. Rep.* **2023**, *50*, 1403–1414. [CrossRef] [PubMed]
22. Huang, W.; Xie, J. Antibacterial effect of dihydromyricetin on specific spoilage organisms of hybrid grouper. *J. Food Qual.* **2021**, *2021*, 5569298. [CrossRef]
23. Markham, K.R. *Technology of Flavonoid Idesntification*; Science Press: Beijing, China, 1990.
24. Robards, K.; Li, X.; Antolovich, M.; Boyd, S. Characterisation of citrus by chromatographic analysis of flavonoids. *J. Sci. Food Agric.* **1997**, *75*, 87–101. [CrossRef]
25. He, X.G. On-line identification of phytochemical constituents in botanical extracts by combined high-performance liquid chromatographic-diode array detection-mass spectrometric techniques. *J. Chromatogr. A* **2000**, *880*, 203–232. [CrossRef]
26. Cuyckens, F.; Shahat, A.A.; Van den Heuvel, H.; Abdel-Shafeek, K.A.; El-Messiry, M.M.; Seif-El Nasr, M.M.; Pieters, L.; Vlietinck, A.J.; Claeys, M. The application of liquid chromatography-electrospray ionization mass spectrometry and collision-induced dissociation in the structural characterization of acylated flavonol O-glycosides from the seeds of *Carrichtera annua*. *Eur. J. Mass Spectrom.* **2003**, *9*, 409–420. [CrossRef]
27. Cavaliere, C.; Cucci, F.; Foglia, P.; Guarino, C.; Samperi, R.; Laganà, A. Flavonoid profile in soybeans by high-performance liquid chromatography/tandem mass spectrometry. *Rapid Commun. Mass. Spectrom.* **2007**, *21*, 2177–2187. [CrossRef]
28. Iwashina, T. Flavonoid function and activity to plants and other organisms. *Biol. Sci. Space* **2003**, *17*, 24–44. [CrossRef] [PubMed]
29. Monici, M.; Baglioni, P.; Mulinacci, N.; Baldi, A.; Vincieri, F. A research model on flavonoids as photoprotectors: Studies on the photochemistry of kaempferol and pelargonidin. *Acta Hortic.* **1993**, *381*, 340–347. [CrossRef]
30. Hamouz, K.; Lachman, J.; Dvorák, P.; Juzl, M.; Pivec, V. The effect of site conditions, variety and fertilization on the content of polyphenols in potato tubers. *Plant Soil. Environ.* **2006**, *52*, 407–412. [CrossRef]
31. Gonçalves, A.F.K.; Friedrich, R.B.; Boligon, A.A.; Piana, M.; Beck, R.C.R.; Athayde, M.L. Anti-oxidant capacity, total phenolic contents and HPLC determination of rutin in *Viola tricolor* (L) flowers. *Free. Radic. Antioxid.* **2012**, *2*, 32–37. [CrossRef]
32. Hertog, M.G.; Kromhout, D.; Aravanis, C.; Blackburn, H.; Buzina, R.; Fidanza, F.; Giampaoli, S.; Jansen, A.; Menotti, A.; Nedeljkovic, S. Flavonoid intake and long-term risk of coronary heart disease and cancer in the seven countries study. *Arch. Intern. Med.* **1995**, *155*, 381–386. [CrossRef] [PubMed]
33. Rimm, E.B.; Katan, M.B.; Ascherio, A.; Stampfer, M.J.; Willett, W.C. Relation between intake of flavonoids and risk for coronary heart disease in male health professionals. *Ann. Intern. Med.* **1996**, *125*, 384–389. [CrossRef] [PubMed]
34. Zheng, S.; Liu, C.; Tang, D.; Zheng, Z.; Yan, R.; Wu, C.; Zuo, N.; Ma, J.; He, Y.; Liu, S. The protective effect of rutin against the cisplatin-induced cochlear damage in vitro. *Neurotoxicology* **2022**, *90*, 102–111. [CrossRef] [PubMed]
35. Pei, H.; Han, C.; Bi, J.; He, Z.; Guo, L. Dihydromyricetin suppresses inflammatory injury in microglial cells to improve neurological behaviors of Alzheimer's disease mice via the TLR4/MD2 signal. *Int. Immunopharmacol.* **2023**, *118*, 110037. [CrossRef]
36. Alamri, Z.Z. Protective and therapeutic effects of apigenin on thioacetamide-induced hepatotoxicity in male rats: Physiological and morphological study. *Egypt. Liver J.* **2024**, *14*, 15. [CrossRef]
37. de Morais, E.F.; de Oliveira, L.Q.R.; Farias Morais, H.G.D.; Souto Medeiros, M.R.D.; Freitas, R.D.A.; Rodini, C.O.; Coletta, R.D. The Anticancer Potential of Kaempferol: A Systematic Review Based on In Vitro Studies. *Cancers* **2024**, *16*, 585. [CrossRef]
38. Deb, B.; Thakur, N. Quercetin Nanoparticles- A Promising Approach to Cancer Treatment: A Review. *Int. J. Life Sci. Pharma Res.* **2022**, *12*, 142–161. [CrossRef]
39. Kumar, S.; Swamy, N.; Tuli, H.S.; Rani, S.; Garg, A.; Mishra, D.; Abdulabbas, H.S.; Sandhu, S.S. Myricetin: A potential plant-derived anticancer bioactive compound—An updated overview. *Naunyn-Schmiedeberg's Arch. Pharmacol.* **2023**, *396*, 2179–2196. [CrossRef]

40. Singla, N.; Gupta, G.; Kulshrestha, R.; Sharma, K.; Bhat, A.A.; Mishra, R.; Patel, N.; Thapa, R.; Ali, H.; Mishra, A.; et al. Daidzein in Traditional Chinese Medicine: A Deep Dive into Its Ethnomedicinal and Therapeutic Applications. *Pharmacol. Res.-Mod. Chin Med.* **2024**, *12*, 100460. [CrossRef]
41. Mangla, B.; Kaur, A.; Kumar, P.; Javed, S.; Ahsan, W.; Popli, H. Development of nanobigel system combining apoptosis-inducing daidzein and flaxseed oil for the synergistic treatment of skin cancer. *J. Drug Deliv. Sci. Technol.* **2024**, *98*, 105843. [CrossRef]
42. Tarahomi, M.; Firouzi Amandi, A.; Eslami, M.; Yazdani, Y.; Salek Farrokhi, A.; Ghorbani, F.; Taherian, M.; Yousefi, B. Niosomes nanoparticles as a novel approach in drug delivery enhances anticancer properties of chrysin in human ovarian carcinoma cells (SKOV3): An in vitro study. *Med. Oncol.* **2023**, *40*, 87. [CrossRef]
43. Li, X.; Li, X.; Deng, L. Chrysin reduces inflammation and oxidative stress and improves ovarian function in D-gal-induced premature ovarian failure. *Bioengineered* **2022**, *13*, 8291–8301. [CrossRef]
44. Shen, J.; Yang, Z.; Wu, X.; Yao, G.; Hou, M. Baicalein facilitates gastric cancer cell apoptosis by triggering endoplasmic reticulum stress via repression of the PI3K/AKT pathway. *Appl. Biol. Chem.* **2023**, *66*, 10. [CrossRef]
45. Zhang, J.; Hu, S.; Gao, Y.; Wei, X.; Qu, Y.; Gao, R.; Lv, Y.; Wang, J.; Wang, Y.; Cao, J.; et al. Galangin alleviated myocardial ischemia-reperfusion injury by enhancing autophagic flux and inhibiting inflammation. *Eur. J. Pharmacol.* **2023**, *945*, 175621 [CrossRef]
46. Park, J.Y.; Park, S.-H.; Oh, S.W.; Kwon, K.; Yu, E.; Choi, S.; Yang, S.; Han, S.B.; Jung, K.; Song, M. Yellow chaste weed and its components, apigenin and galangin, affect proliferation and oxidative stress in blue light-irradiated HaCaT cells. *Nutrients* **2022** *14*, 1217. [CrossRef]
47. Nabizadeh, Z.; Nasrollahzadeh, M.; Shabani, A.A.; Mirmohammadkhani, M.; Nasrabadi, D. Evaluation of the anti-inflammatory activity of fisetin-loaded nanoparticles in an in vitro model of osteoarthritis. *Sci. Rep.* **2023**, *13*, 15494. [CrossRef]
48. Mustafa, S.; Anwar, H.; Ain Qu Ahmed, H.; Iqbal, S.; Ijaz, M.U. Therapeutic effect of gossypetin against paraquat-induced testicular damage in male rats: A histological and biochemical study. *Environ. Sci. Pollut. Res.* **2023**, *30*, 62237–62248. [CrossRef]
49. Rumman, M.; Pandey, S.; Singh, B.; Gupta, M.; Ubaid, S.; Mahdi, A.A. Genistein prevents hypoxia-induced cognitive dysfunctions by ameliorating oxidative stress and inflammation in the hippocampus. *Neurotox. Res.* **2021**, *39*, 1123–1133. [CrossRef]
50. Kang, Z.C.; Wang, H.G.; Yang, Y.L.; Zhao, X.Y.; Zhou, Q.M.; Yang, Y.L.; Yang, J.Y.; Du, G.H. Pinocembrin ameliorates cognitive impairment induced by vascular dementia: Contribution of Reelin-dab1 signaling pathway. *Drug Des. Dev. Ther.* **2020**, *14*, 3577–3587. [CrossRef]
51. Zou, Z.C.; Fu, J.J.; Dang, Y.Y.; Zhang, Q.; Wang, X.F.; Chen, H.B.; Jia, X.J.; Lee, S.M.-Y.; Li, C.W. Pinocembrin-7-methylether protects SH-SY5Y cells against 6-hydroxydopamine-induced neurotoxicity via modulating Nrf2 induction through AKT and ERK pathways. *Neurotox. Res.* **2021**, *39*, 1323–1337. [CrossRef]
52. August, P.M.; Santos, B.G.D. Naringin and naringenin in neuroprotection and oxidative stress. In *Oxidative Stress and Dietary Antioxidants in Neurological Diseases*; Academic Press: New York, NY, USA, 2020. [CrossRef]
53. Huang, H.; He, Y.; Li, Y.; Gu, M.; Wu, M.; Ji, L. Eriodictyol suppresses the malignant progression of colorectal cancer by downregulating tissue specific transplantation antigen P35B (TSTA3) expression to restrain fucosylation. *Bioengineered* **2022**, *13*, 5551–5563. [CrossRef]
54. Zhu, Z.; Yang, L.; Li, Z.; Liu, Q. Cyanidin-3-O-glucoside, cyanidin, and oxidation products of cyanidin protect neuronal function through alleviating inflammation and oxidative damage. *J. Food Sci.* **2022**, *87*, 2159–2172. [CrossRef]
55. Song, M.-T.; Wang, W.-Z.; Lu, Y.; Han, R.-M.; Skibsted, L.H.; Zhang, J.-P. Double-Site Binding and Anti-/Pro-oxidation of Luteolin on Bovine Serum Albumin Mediated by Copper (II) Coordination. *ACS Omega* **2022**, *7*, 19521–19534. [CrossRef]
56. Tan, X.; Yang, Y.; Xu, J.; Zhang, P.; Deng, R.; Mao, Y.; He, J.; Chen, Y.; Zhang, Y.; Ding, J. Luteolin exerts neuroprotection via modulation of the p62/Keap1/Nrf2 pathway in intracerebral hemorrhage. *ACS Omega* **2020**, *10*, 1551. [CrossRef]
57. Patel, D.K. Potential Benefits of Tricetin in Medicine for the Treatment of Cancers and Other Health-Related Disorders: Medicinal Importance and Therapeutic Benefit. *Nat. Prod. J.* **2022**, *12*, 12–19. [CrossRef]
58. Wei, A.; Zhu, H.; Xie, F.; Zhang, M.; Huang, P.; Yang, W. Advances in the Treatment of Hepatitis with Flavonoids. *Med. Plant* **2021**, *12*, 70–74.
59. Hu, Q.; Qu, C.; Xiao, X.; Zhang, W.; Jiang, Y.; Wu, Z.; Song, D.; Peng, X.; Ma, X.; Zhao, Y. Flavonoids on diabetic nephropathy: Advances and therapeutic opportunities. *Chin. Med.* **2021**, *16*, 74. [CrossRef]
60. Gupta, A.K.; Kumar, S. Review on Diabetic Complications and their Management by Flavonoids and Triterpenoids. *Nat. Prod. J.* **2023**, *13*, 105–114.
61. Parashar, D.; Meena, A.K.; Sharma, M. Flavonoids—A Review on Natural Antioxidants. *Res. J. Pharm. Technol.* **2023**, *16*, 4952–4958. [CrossRef]
62. William, N.; Lala, G.; Ethan, J.; Caitlyn, S.; Dalia, S. Review Article: The in vivo Anti-Candida albicans Activity of Flavonoids. *J. Oral. Biosci.* **2021**, *63*, 120–128.
63. Wei, Q.; Zhang, Y.H. Flavonoids with Anti-Angiogenesis Function in Cancer. *Molecules* **2024**, *29*, 1570–1601. [CrossRef]
64. Vetrivel, P.; Kim, S.M.; Saralamma, V.V.G.; Ha, S.E.; Kim, E.H.; Min, T.S.; Kim, G.S. Function of flavonoids on different types of programmed cell death and its mechanism: A review. *J. Biomed. Res.* **2019**, *33*, 363. [CrossRef]
65. Khazdair, M.R.; Anaeigoudari, A.; Agbor, G.A. Anti-viral and anti-inflammatory effects of kaempferol and quercetin and COVID-2019: A scoping review. *Asian Pac. J. Trop. Biomed.* **2021**, *11*, 327–334. [CrossRef]

66. Khan, J.; Deb, P.K.; Priya, S.; Medina, K.D.; Devi, R.; Walode, S.G.; Rudrapal, M. Dietary flavonoids: Cardioprotective potential with antioxidant effects and their pharmacokinetic, toxicological and therapeutic concerns. *Molecules* **2021**, *26*, 4021. [CrossRef]
67. Zhang, J.J.; Shao, M.M.; Wang, M.C. Therapeutic potential of natural flavonoids in pulmonary arterial hypertension: A review. *Phytomed* **2024**, *128*, 155535. [CrossRef]
68. Sapra, K.; Sharma, R.; Samamadar, S.; Sapra, A.; Kumar, V. Current Trends and Prospective Role of Flavonoids and Flavones as Anticancer Agents: A Review. *Curr. Nutr. Food Sci.* **2023**, *19*, 712–722.
69. De Sousa Silva, G.V.; Lopes, A.L.V.F.G.; Viali, I.C.; Lima, L.Z.M.; Bizuti, M.R.; Haag, F.B.; Tavares de Resende e Silva, D. Therapeutic properties of flavonoids in treatment of cancer through autophagic modulation: A systematic review. *Chin. J. Integr. Med.* **2023**, *29*, 268–279. [CrossRef]
70. Abusaliya, A.; Ha, S.E.; Bhosale, P.B.; Kim, H.H.; Park, M.Y.; Vetrivel, P.; Kim, G.S. Glycosidic flavonoids and their potential applications in cancer research: A review. *Mol. Cell. Toxicol.* **2022**, *18*, 9–16. [CrossRef]
71. Hamsalakshmi; Alex, A.M.; ArehallyMarappa, M.; Joghee, S.; Chidambaram, S.B. Therapeutic benefits of flavonoids against neuroinflammation: A systematic review. *Inflammopharmacology* **2022**, *30*, 111–136. [CrossRef]
72. Munguía, L.; Ortiz, M.; González, C.; Portilla, A.; Meaney, E.; Villarreal, F.; Najera, N.; Ceballos, G. Beneficial Effects of Flavonoids on Skeletal Muscle Health: A Systematic Review and Meta-Analysis. *J. Med. Food* **2022**, *25*, 465–486. [CrossRef]
73. Cao, Y.; Xie, L.; Liu, K.; Liang, Y.; Dai, X.; Wang, X.; Lu, J.; Zhang, X.; Li, X. The antihypertensive potential of flavonoids from Chinese Herbal Medicine: A review. *Pharmacol. Res.* **2021**, *174*, 105919. [CrossRef]
74. Rufino, A.T.; Costa, V.M.; Carvalho, F.; Fernandes, E. Flavonoids as antiobesity agents: A review. *Med. Res. Rev.* **2021**, *41*, 556–585. [CrossRef]
75. Yao, J.; Zhang, Y.; Wang, X.Z.; Zhao, J.; Yang, Z.J.; Lin, Y.P.; Sun, L.; Lu, Q.Y.; Fan, G.J. Flavonoids for treating viral acute respiratory tract infections: A systematic review and meta-analysis of 30 randomized controlled trials. *Front. Public Health* **2022**, *10*, 814669. [CrossRef]
76. Ali, S.; Corbi, G.; Maes, M.; Scapagnini, G.; Davinelli, S. Exploring the impact of flavonoids on symptoms of depression: A systematic review and meta-analysis. *Antioxidants* **2021**, *10*, 1644. [CrossRef]
77. Khater, M.; Greco, F.; Osborn, H.M.I. Antiangiogenic Activity of Flavonoids: A Systematic Review and Meta-Analysis. *Molecules* **2020**, *25*, 4712. [CrossRef]
78. Spoerlein, C.; Mahal, K.; Schmidt, H.; Schobert, R. Effects of chrysin, apigenin, genistein and their homoleptic copper (II) complexes on the growth and metastatic potential of cancer cells. *J. Inorg. Biochem.* **2013**, *127*, 107–115. [CrossRef]
79. Xiao, M.F.; Zhao, B.Q.; Liu, F.; Wu, P. Optimization of the extraction process of chrysin-7-glucoside from Sanyechong tea. *Cent. South Pharm.* **2014**, *12*, 321–324.
80. Zhou, Q.; Shen, D.; Qu, Y. Microwave-assisted Extraction of Daidzein from Defatted Soybean Meal after Extrusion. *Soy. Sci.* **2008**, *27*, 124.
81. Yang, Y.C.; Wei, M.C.; Huang, T.C.; Lee, S.Z.; Lin, S.S. Comparison of modified ultrasound-assisted and traditional extraction methods for the extraction of baicalin and baicalein from *Radix Scutellariae*. *Ind. Crops Prod.* **2013**, *45*, 182–190. [CrossRef]
82. Ong, E.S.; Len, S.M. Pressurized hot water extraction of berberine, baicalein and glycyrrhizin in medicinal plants. *Anal. Chim. Acta* **2003**, *482*, 81–89. [CrossRef]
83. Liu, Y.H.; Huang, Z.F.; Chen, Y.; Liu, Y.H.; Jin-hai, Y. Extraction of Baicalein in *Scutellaria baicalensis* by Enzymic Dissolution. *Nat. Prod. Res. Dev.* **2007**, *19*, 688–691.
84. Kim, H.H.; Bae, Y.; Kim, S.H. Galangin attenuates mast cell-mediated allergic inflammation. *Food Chem. Toxicol.* **2013**, *57*, 209–216. [CrossRef]
85. Kang, A.; Yan, M.; Peng, Y. Study on ethonal-extraction and purification process of galangin from *Alpinia officinarum* hance. *Chin. J. Inform. Tradit. Chin. Med.* **2011**, *18*, 44–46.
86. Fu, Y.J.; Liu, W.; Zu, Y.G.; Tong, M.H.; Li, S.M.; Yan, M.M.; Efferth, T.; Luo, H. Enzyme assisted extraction of luteolin and apigenin from pigeonpea [*Cajanuscajan* (L.) Millsp.] leaves. *Food Chem.* **2008**, *111*, 508–512. [CrossRef]
87. Huang, F.Y.; Li, G.P.; Wu, J.J.; Tang, R.P.; Zhang, Y.H. The Process research of extraction of apigenin in *Apium graveolens*. *Guangzhou Chem. Ind.* **2013**, *41*, 133–148.
88. Olsen, H.T.; Stafford, G.I.; Van Staden, J.; Christensen, S.B.; Jäger, A.K. Isolation of the MAO-inhibitor naringenin from *Mentha aquatica* L. *J. Ethnopharmacol.* **2008**, *117*, 500–502. [CrossRef]
89. Zhang, D.Y.; Zu, Y.G.; Fu, Y.J.; Luo, M.; Gu, C.B.; Wang, W.; Yao, X.H. Negative pressure cavitation extraction and antioxidant activity of biochanin A and genistein from the leaves of *Dalbergia odorifera* T. Chen. *Sep. Purif. Technol.* **2011**, *83*, 91–99. [CrossRef]
90. Fu, Y.; Zu, Y.; Liu, W.; Efferth, T.; Zhang, N.; Liu, X.; Kong, Y. Optimization of luteolin separation from pigeonpea [*Cajanus cajan* (L.) Millsp.] leaves by macroporous resins. *J. Chromatogr. A* **2006**, *1137*, 145–152. [CrossRef]
91. Tang, F. Extracting luteolin from *Callicarpa formosana* Rolfe. *China J. Mod. Med.* **2014**, *24*, 75–77.
92. Li, J.H.; Duan, D.D.; Ma, C.L.; Shen, Q.L. Fisetin extraction process of *cotinus coggygria*'s active ingredients. *China J. Tradit. Chin. Med. Pharm.* **2012**, *27*, 1956–1958.
93. Sladkovský, R.; Solich, P.; Opletal, L. Simultaneous determination of quercetin, kaempferol and (E)-cinnamic acid in vegetative organs of *Schisandra chinensis* Baill. by HPLC. *J. Pharm. Biomed. Anal.* **2001**, *24*, 1049–1054. [CrossRef] [PubMed]
94. Bo, W.; Wei-nong, Z.; Jian-fang, Y.; Han-jun, Z. Optimum Extraction Technology of Kaempferol in *Prunus cerasifera* Leaves. *Nat. Prod. Res. Dev.* **2011**, *23*, 952–955.

95. Zhang, Y.J.; Liu, Z.M.; Xu, Z.G. Determination of quercetin in *Erigeron breviscapus* by high performance liquid chromatograp *Chem. Res.* **2014**, *25*, 140–143.
96. Harbourne, N.; Jacquier, J.C.; O'Riordan, D. Optimisation of the aqueous extraction conditions of phenols from meadowsw (*Filipendula ulmaria* L.) for incorporation into beverages. *Food Chem.* **2009**, *116*, 722–727. [CrossRef]
97. Vuong, Q.V.; Golding, J.B.; Stathopoulos, C.E.; Nguyen, M.H.; Roach, P.D. Optimizing conditions for the extraction of catechi from green tea using hot water. *J. Sep. Sci.* **2011**, *34*, 3099–3106. [CrossRef]
98. Zhang, H.; Liu, M.; Han, S.; Wei, Y. Optimizing the extraction of catechin from peanut red skin using response surface methodolo and its antioxidant activity. *IERI Procedia* **2013**, *5*, 312–320. [CrossRef]
99. Luo, Z.Y.; Yang, X.P.; Wu, M.C. Studies on the Extraction of Water- soluble Polysaccharides and Total Flavone from *Ampelops grossedentata*. *Food Sci.* **2005**, *26*, 156–160.
100. Qin, J.P.; Xu, X.J.; Dong, M.J. Study on the new extraction method of the flavonoids in *Ampelopsis grossedentata*. *Chin. J. Mod. Ap Pharm.* **2000**, *17*, 196–197.
101. Yao, Q.; Guo, X.Q.; Song, Q.; Wu, X.Y.; Zhang, L.L.; Lu, J.; Gou, X.J. Comparison of rutin extraction technologies from Huain *J. Chengdu Univ. (Nat. Sci. Ed.)* **2013**, *32*, 332–334.
102. Huang, R.H.; Han, W.; Zhou, Y.C.; Li, Y.; Deng, X. Pilot experiment of microwave-assisted extraction of icariin. *Chin. Tradit. P Med.* **2004**, *26*, 525–528.
103. Wang, Y.J.; Yu, H.R.; Wang, H.X.; Yang, J. Comparative studies on the extraction and purification of baicalin. *Chin. J. New Dru* **2007**, *16*, 1101–1104.
104. Andersen, O.M.; Markham, K.R. *Flavonoids: Chemistry, Biochemistry and Applications*; CRC Press: Boca Raton, FL, USA, 2005.
105. Yang, X.H.; Lv, B.Y.; Mao, Q.L. Study on alkali extraction and acid precipitation of rutin from bud of *Sophora japonica*. *Sci. Techn Food Ind.* **2007**, *28*, 183–185.
106. Vinitha, U.G.; Sathasivam, R.; Muthuraman, M.S. Intensification of supercritical fluid in the extraction of flavonoids: A compr hensive review. *Physiol. Mol. Plant Pathol.* **2022**, *118*, 101815. [CrossRef]
107. Mujwah, A.A.; Mohammed, M.A.; Ahmed, M.H. First isolation of a flavonoid from *Juniperus procera* using ethyl acetate extrac *Arab. J. Chem.* **2010**, *3*, 85–88. [CrossRef]
108. Hemwimol, S.; Pavasant, P.; Shotipruk, A. Ultrasound-assisted extraction of anthraquinones from roots of *Morinda citrifoli Ultrason. Sonochem.* **2006**, *13*, 543–548. [CrossRef]
109. Zhang, H.F.; Yang, X.H.; Zhao, L.D.; Wang, Y. Ultrasonic-assisted extraction of epimedin C from fresh leaves of Epimedium an extraction mechanism. *Innov. Food Sci. Emerg. Technol.* **2009**, *10*, 54–60. [CrossRef]
110. Dong, J.; Gao, W.; Lu, D.; Wang, Y. Simultaneous extraction and analysis of four polyphenols from leaves of *Lycium barbarum* I *J. Food Biochem.* **2011**, *35*, 914–931. [CrossRef]
111. Zhang, H.F.; Zhang, X.; Yang, X.H.; Qiu, N.X.; Wang, Y.; Wang, Z.Z. Microwave assisted extraction of flavonoids from cultivate *Epimedium sagittatum*: Extraction yield and mechanism, antioxidant activity and chemical composition. *Ind. Crops Prod.* **2013**, *5* 857–865. [CrossRef]
112. Tu, Z.C.; Zhang, L.; Wang, H. Effects of Dynamic High Pressure Microfluidization (DHPM) Extraction on the Antioxidation o Flavonoids from Sweet Potato Leaves and Mechanism Studying. *Chin. J. High Press. Phys.* **2013**, *27*, 431–438.
113. Ru, Q.; Jin, M.; Su, Y.; Shen, J.F. Research progress on synergistic effects of natural flavonoids. *Nat. Prod. Res. Dev.* **2023**, *35*, 531–538
114. Chen, Q.P.; Lin, X.J.; Li, S.L.; Liu, H.S.; Qi, X.Y. Synergistic Anti-tumor Effect of Dihydromyricetin and Myricitrin on HepG2 cell *Sci. Tech. Food Ind.* **2021**, *8*, 333–337.
115. De Rijke, E.; Out, P.; Niessen, W.M.A.; Ariese, F.; Gooijer, C.; Brinkman, U.A.T. Analytical separation and detection methods fo flavonoids. *J. Chromatogr. A* **2006**, *1112*, 31–63. [CrossRef] [PubMed]
116. Robards, K. Strategies for the determination of bioactive phenols in plants, fruit and vegetables. *J. Chromatogr. A* **2003**, *100* 657–691. [CrossRef]
117. Rolim, A.; Oishi, T.; Maciel, C.P.M.; Zague, V.; Pinto, C.; Kaneko, T.M.; Consiglieri, V.O.; Velasco, M.V.R. Total flavonoids quantification from O/W emulsion with extract of Brazilian plants. *Int. J. Pharm. X* **2006**, *308*, 107–114. [CrossRef] [PubMed]
118. Dong, J.; Wang, Y. Quantitation and identification of the flavonoids present in fruits of *Lycium barbarum*. *Food Res. Dev.* **2009**, *30*, 36–40
119. Lei, C.; Ma, Q.; Tang, Q.Y.; Ai, X.R.; Zhou, Z.; Yao, L.; Wang, Y.; Wang, Q.; Dong, J.Z. Sodium selenite regulates phenolics accumulation and tuber development of purple potatoes. *Sci. Hortic.* **2014**, *165*, 142–147. [CrossRef]
120. Nguyen, T.H.; Vu, D.C.; Ngan, N.T.; Tran-Trung, H.; Dang, V.S. Determination of Phenolics and Biological Activity of *Viola dalatensis* Gagnep. by High-Performance Liquid Chromatography (HPLC) and Multivariate Analysis. *Anal. Lett.* **2024**, *57*, 1666–1677. [CrossRef]
121. Longkumer, I.; Chaturvedi, R. Determination of Rutin in Musa Leaves by High Performance Liquid Chromatography—A Step Towards Sustainability. *In Vitro Cell. Dev. Biol. Plant.* **2022**, *58*, 698.
122. Bhagat, S.; Rathore, M.; Kachhwaha, S.; Sharma, H.K. Phytochemical screening, determination of Total phenol content, Total flavonoid content and quantitative estimation of Rutin and quercetin using RP-HPLC in the fruits of *Capparis decidua* (Forsk.). *Ind. J. Pure App. Ind. J. Pure App. Biosci.* **2021**, *9*, 254–261. [CrossRef]

123. Nguyen, T.; Nguyen, T.; Dam, T. Simultaneous determination of Rutin, Hesperidin and Quercetin contents in solid dietary supplements by high performance liquid chromatography. *Vietnam. J. Food Control* **2020**, *3*, 38–45. [CrossRef]
124. Binang, K.; Takuwa, D.T. Development of reverse phase-high performance liquid chromatography (RP-HPLC) method for determination of selected antihypertensive active flavonoids (rutin, myricetin, quercetin, and kaempferol) in medicinal plants found in Botswana. *Phys. Sci. Rev.* **2023**, *8*, 679–692. [CrossRef]

MDPI AG
Grosspeteranlage 5
4052 Basel
Switzerland
Tel.: +41 61 683 77 34

Separations Editorial Office
E-mail: separations@mdpi.com
www.mdpi.com/journal/separations

www.ingramcontent.com/pod-product-compliance
Lightning Source LLC
Chambersburg PA
CBHW062053220926
51850CB00010B/2750